T0335514

The Statistical Foundations of Entropy

The Statistical Foundations of Entropy

John D Ramshaw
Portland State University, USA

World Scientific

NEW JERSEY • LONDON • SINGAPORE • BEIJING • SHANGHAI • HONG KONG • TAIPEI • CHENNAI • TOKYO

Published by

World Scientific Publishing Co. Pte. Ltd.
5 Toh Tuck Link, Singapore 596224
USA office: 27 Warren Street, Suite 401-402, Hackensack, NJ 07601
UK office: 57 Shelton Street, Covent Garden, London WC2H 9HE

British Library Cataloguing-in-Publication Data
A catalogue record for this book is available from the British Library.

THE STATISTICAL FOUNDATIONS OF ENTROPY

ISBN 978-981-3234-12-3

For any available supplementary material, please visit
http://www.worldscientific.com/worldscibooks/10.1142/10823#t=suppl

Printed in Singapore

In Memoriam

J. Willard Gibbs

and

Edwin T. Jaynes

"$S = \log W$, $W = \exp S$," – that is all
Ye know on earth, and all ye need to know.

PREFACE

This book presents a unified but somewhat unorthodox account of the statistical foundations of entropy and the fundamentals of equilibrium statistical mechanics. The historical evolution of these intimately related subjects was inevitably somewhat haphazard, but with the passage of time they can now be seen to comprise a cohesive conceptual and logical structure. However, the simplicity, consistency, and coherence of that structure are often obscured by the fragmented historical manner in which these subjects have traditionally been taught and treated in textbooks. This book is the result of a systematic effort to reassess and reorganize this fundamental material into a simpler hierarchial framework in which each level of the logical structure naturally follows from and builds upon the preceding levels. In this way the sequence in which the various topics are discussed accurately reflects their logical interdependencies, which are not always obvious in more traditional treatments. It seems to me that the resulting progressive organization of this book provides a significantly shorter, simpler, and more straightforward route to a clear understanding of the essentials of entropy and statistical mechanics.

The Table of Contents also serves as an outline of the book, but does not adequately emphasize its main distinguishing features, which are as follows:

• The entire development is predicated on the fundamental Boltzmann-Planck and Boltzmann-Gibbs-Shannon expressions $S =$

$\log W$ and $S = -\sum_k p_k \log p_k$ for the entropy S, which are shown to be immediate and nearly unique consequences of interpreting S as a consistent additive measure of the uncertainty as to which of its accessible states a system occupies. These expressions are not restricted to thermodynamic or many-particle systems.

• The ambiguity in extending the definition of entropy to systems with continuous states, which has been a source of considerable confusion, is formally resolved simply by introducing the density of discrete states in a continuous state space. This transfers the ambiguity to the density of states, where it is more easily dealt with.

• Several important concepts are discussed in more depth and detail than usual, including the hypothesis of equal *a priori* probabilities, the general treatment of constraints, the principle of maximum entropy, and the statistical interpretation of thermodynamics. Special attention is given to the important but oft-neglected requirement that the formulation be properly covariant with respect to arbitrary nonlinear transformations of the state variables.

• In contrast to most introductory treatments, many-particle systems are not discussed until the second half of the book. This reflects and emphasizes the fact that in a logical sense such systems constitute a special case (albeit the most important one) of the general formalism, most of which is entirely independent of the structure and enumeration of many-particle microstates.

• The tedious combinatorial arguments which play a prominent role in many introductory treatments are reduced to the bare minimum required to characterize the states of many-particle systems in terms of multinomial coefficients. This de-emphasis reflects my view that combinatorial arguments are in essence mere technical tools used to enumerate many-particle microstates, and as such are only peripherally related to the conceptual structure of entropy per se.

• The vast majority of textbooks and treatises on statistical mechanics perpetuate the persistent misconception that the indistinguishability of identical particles derives from quantum mechanics, and that identical classical particles are distinguishable. In contrast, the present development is firmly and unequivocally based on the proposition that identical particles are inherently indistinguishable in both classical and quantum mechanics. This principle unifies and simplifies the logical structure of the subject by eliminating the artificial dichotomy between classical and quantum statistics which afflicts conventional treatments. The Gibbs paradox then never arises, and its associated factor of $1/N!$ appears automatically and need not be introduced as an *ad hoc* correction factor.

As the above summary and the Table of Contents imply, the scope of this book is restricted to fundamental concepts. With the sole exception of ideal gases, the many important applications of equilibrium statistical mechanics to interpret and predict the properties of matter are entirely omitted. Detailed discussions of such applications are of course readily available in other books, including those listed in the Bibliography. Neither are problems or exercises provided to assist readers in assessing their understanding of the material. For these reasons, this book is neither suitable nor intended for use as a primary textbook for a first course in statistical mechanics. However, I hope it will be found useful as a supplementary reference for courses in statistical mechanics, thermal physics, thermodynamics, and physical chemistry, especially by students seeking a deeper or more detailed discussion of the fundamental concepts. This book may also be suitable for use in second, advanced, or special topics courses in which the fundamentals of the subject are critically reexamined. Its sequential organization should also make it well suited for self-study.

This book is written at a level which should be accessible to advanced undergraduate and beginning graduate students in physics, engineering, physical chemistry, applied mathematics, and related fields. Such students are presumed to be conversant with the rudiments of probability theory, including random variables, joint and conditional probability densities and distributions, mean and most probable values, fluctuations, variance, and statistical independence. Chapters 6–8 presuppose familiarity with the fundamentals and mathematical manipulations of classical thermodynamics. However, the discussion is relatively concise and the number density of concepts per page is rather high in some places, so the book is unlikely to reward a casual perusal. *Au contraire*, it requires concentration and careful attention to detail, so readers encountering the statistical interpretation of entropy for the first time are likely to benefit from frequent pauses for thoughtful contemplation. I hope those who persevere will find the investment worthwhile, and will thereby acquire a deeper appreciation for the subtle beauty, elusive simplicity, and ultimate inevitability of entropy.

Acknowledgments

I am indebted to Ralph F. Baierlein, Erik Bodegom, Andrew W. Cook, Glen A. Hansen, and John F. Nagle for kindly taking the time and trouble to read parts or all of the manuscript and offer helpful comments, advice, suggestions, criticisms, and/or corrections. Their assistance is greatly appreciated and gratefully acknowledged, and they are hereby absolved of any responsibility for whatever errors, inaccuracies, misconceptions, or other deficiencies may remain, which I only hope are logarithmically negligible.

CONTENTS

RELATIONAL NOTATION

$\equiv$ means "is equal to by definition"

$\cong$ means "is approximately equal to"

$\sim$ means "is of the same order of magnitude as," and does *not* imply asymptotic equality

$\lesssim$ means "is less than or approximately equal to"

$\gtrsim$ means "is greater than or approximately equal to"

Chapter 1

INTRODUCTION

Entropy occupies a secure and prominent position on any list of the most fundamental and important concepts of physics, but this trite observation does not fully do it justice. Entropy has a certain solitary mystique all its own, which both fascinates and frustrates those who aspire to comprehend its elusive and multifaceted subtleties. Unlike mass, momentum, and energy, the entropy of an isolated system is not conserved but has the peculiar property of spontaneously increasing to a maximum. A quantity which can only grow larger is curious if not unsettling to contemplate; it vaguely suggests instability, or even some inexorable impending catastrophe worthy of a science fiction movie, perhaps entitled *The Heat Death of the Universe!* It also seems curious that in spite of its thermodynamic origins, entropy cannot be fully understood in thermodynamic terms alone but requires statistical concepts for its complete elucidation. Indeed, the statistical interpretation of entropy is in many respects simpler and easier to comprehend than its thermodynamic aspects, and arguably provides the most transparent pedagogical approach to the subject. This of course is the rationale for treating thermodynamics and statistical mechanics together as a single discipline, an approach which was once considered heretical but has now become accepted and commonplace.

The statistical interpretation of entropy is approaching its sesquicentennial, and by now has been explored and expounded in a virtually uncountable and unsurveyable multitude of papers, textbooks, treatises, essays, reviews, conference proceedings, etc. This prolonged collective endeavor has largely demystified entropy, while at the same time expanding its scope well beyond the confines of thermodynamics. Indeed, entropy has become perhaps the most pervasive and multifaceted of all physical concepts in terms of its wide applicability to a variety of other disciplines. In short, entropy emerged from its formative years long ago and has now evolved into a mature concept which for the most part is well established and well understood. In spite of this, however, the concept of entropy has yet to converge to a stable equilibrium state. Although the correct formulae to be employed in applications are by now well established, there is no similar consensus as to their proper derivation and interpretation. Moreover, the interpretation, significance, and even the proper definition of entropy itself remain remarkably controversial and contentious [1–8]. This situation is rather disconcerting and can hardly be regarded as satisfactory.

This book represents an attempt to ameliorate the residual controversy and conceptual confusion surrounding entropy by revisiting and systematically reconstructing its statistical foundations *ab initio*, making full use of the clarity of hindsight. Our intention and objective has been to develop and present those foundations and their most basic implications in a straightforward, concise, unified, and hopefully cogent form. To this end, the logical structure of the subject is emphasized by clearly stating the minimal postulates and hypotheses required in order to proceed, and deriving their logical consequences as simply and economically as possible. The development is therefore essentially axiomatic and deductive in spirit and structure, but in an informal physical sense rather than a rigorous mathematical sense. An assiduous effort has been made to avoid logical lacunae and non sequiturs, which alas are not uncommon in textbooks. As described in the Preface, the material has been organized in the manner which seemed to result in the most direct and transparent logical structure, which in several respects departs significantly from the orthodox beaten path. The overall organization and main distinguishing features of the book are indicated in the Preface and Table of Contents, so it would be redundant to reiterate that information here. The remainder of this introductory discussion accordingly focuses on those aspects of the treatment which warrant further emphasis or elaboration at the outset.

The present development is predicated on the deceptively simple but highly fruitful proposition that the essence of entropy as a statistical concept is that it represents a consistent quantitative measure of uncertainty which is additive for statistically independent systems. Somewhat surprisingly, this sole and superficially trivial criterion suffices to determine the entropy almost uniquely. The remainder of the development consists, at least in principle, mainly in mere details of elaboration, but as usual the latter are wherein the Devil resides. Indeed, it would not be entirely inaccurate to say that it's all uphill from there, or that entropy is so simple that it borders on the incomprehensible, in the sense that it is at first difficult to imagine how such an apparently simplistic notion could have such subtle and profound consequences. Thus, to contraphrase Einstein, our task is to make statistical entropy as complicated as necessary to be comprehensible, but hopefully no more so.

Uncertainty implies a plurality of possibilities, which in turn implies the need for a statistical or probabilistic description. For example, the uncertainty as to whether flipping a coin will result in heads (H) or tails (T) evidently depends on the relative probabilities $p(\mathrm{H})$ and $p(\mathrm{T}) = 1 - p(\mathrm{H})$ of those two outcomes, and is clearly larger if the coin is fair (i.e., $p(\mathrm{H}) = p(\mathrm{T}) = 1/2$) than if the coin is loaded so that the outcome is almost certain to be heads (e.g., $p(\mathrm{H}) = 1 - p(\mathrm{T}) = 0.99$). This observation suggests that the uncertainty of a situation is largest, *ceteris paribus*, when all possibilities are equally likely. Conversely, in situations where the uncertainty is perceived to be a maximum it is therefore not unreasonable to presume, at least provisionally, that all outcomes are equally likely. This presumption is traditionally referred to as the hypothesis of equal *a priori* probabilities (EAPP). This hypothesis is obviously inappropriate in situations where the probabilities are manifestly unequal, so it must be expected that its validity will in general be subject to certain restrictions. In physical applications, the most important such restriction is to systems of constant energy. What is remarkable is that virtually all of equilibrium statistical mechanics can then be derived, in a relatively straightforward way, by combining the EAPP hypothesis with the superficially imprecise conception of entropy as a quantitative measure of uncertainty.

The most important applications of entropy are to composite systems comprising a very large number of elementary parts or components which can be arranged in a very large number of possible ways or configurations. Those configurations constitute the states of the system, and the uncertainty then pertains to which of those states the system occupies. Those states are almost always defined and characterized in terms of the states of the individual components. As a trivial example, the states of an individual coin are H and T, so a system of two distinguishable coins has four possible states, which can be represented by the ordered pairs (H, H), (H, T), (T, H), and (T, T). In contrast, if the coins are identical and indistinguishable there are only three possible states: $\{\mathrm{H}, \mathrm{H}\}$ (both heads), $\{\mathrm{T}, \mathrm{T}\}$ (both tails), and $\{\mathrm{H}, \mathrm{T}\} = \{\mathrm{T}, \mathrm{H}\}$ (one of each).

The composite systems of greatest interest in physics are macroscopic thermodynamic systems, which are composed of very large numbers of very small particles (usually atoms, molecules, ions, and/or free electrons). For simplicity we shall restrict attention to pure systems whose constituent particles are all of the same type or species. Mixtures are therefore specifically excluded from the discussion. They present no special conceptual difficulties, but they introduce cumbersome notational and bookkeeping complications and distractions which are best avoided in a critical discussion of the fundamentals. The omission of mixtures implies that our primary focus is on systems of identical indistinguishable particles. However, it is necessary to consider distinguishable particles as well, because a clear conceptual understanding of the one ironically requires an understanding of the other. Indeed, as will be seen, the states of a system of distinguishable particles are more easily defined and characterized than those of an otherwise identical system of indistinguishable particles, so it is sometimes useful to describe the latter in terms of the former. However, mathematical convenience must not be confused with physical reality. If a physical system is composed of physically indistinguishable particles, then the corresponding system of distinguishable particles is fictitious, and so are its states, however useful they may be mathematically.

As noted in the Preface, this book differs from many if not most introductory treatments of statistical mechanics in that (1) many-particle systems are not discussed until relatively late in the book, and (2) the combinatorial arguments employed for that purpose are minimized. The reasons for these deviations from the norm are as follows: (1) Much of the general formalism of entropy is actually independent of the detailed structure of the states and the variables used to describe them. This generality is obscured if the formalism is specialized to particular cases such as many-particle systems before it is necessary to do so. (2) More conventional treatments in which combinatorial concepts play a central role exaggerate their importance in the conceptual structure of entropy. Indeed, such concepts do not even arise until the general formalism is applied to composite systems, where they play a limited albeit important role as mathematical tools used to enumerate and classify the states. Even for that purpose, they can be largely circumvented by the use of simpler and more elegant analytical techniques, in particular the canonical and grand canonical probability distributions. Although statistical mechanics can be, and often still is, motivated and developed by means of combinatorial arguments, they do not in my judgment provide the simplest or most transparent approach to the subject. Of course, such arguments played an important role in its historical development, of which the best known and most important example is the conceptually precarious "method of the most probable distribution" developed by Boltzmann circa 1877. This method is still commonly invoked in introductory treatments because of its simplicity, even though it is universally regarded and acknowledged as unsatisfactory. In contrast, I regard it as being of historical interest only and make no use of it in this book.

Perhaps the most heretical distinguishing feature of the present treatment is its categorical rejection of the persistent misconception that the indistinguishability of identical particles is peculiar to quantum mechanics, and that identical classical particles are in principle distinguishable. This unfortunate misconception has afflicted statistical mechanics for nearly a century, and its eviction from the subject is long overdue. It has been propagated from each generation of textbooks to the next for so long that it has evolved into a pernicious memetic defect which most authors simply take for granted and no longer subject to critical scrutiny. As a result, the alleged distinguishability of identical classical particles remains firmly entrenched as established dogma, a situation which future historians of science are likely to find incomprehensible. However, this misconception has not gone entirely unchallenged; it has been intermittently disputed and/or refuted on various occasions by several intrepid authors [2,9–16], not to mention Gibbs himself [17], but from various different perspectives which have quite likely detracted from their collective cogency. In particular, some authors seem to consider indistinguishability a property of the states rather than the particles, whereas this book is based on the converse view.

The present treatment of multi-particle systems is firmly based on the principle that *indistinguishability is an intrinsic property of identical particles in general, and is not peculiar to quantum mechanics*. Readers who are unfavorably disposed toward that principle are encouraged to peruse Chapter 9, where we adduce arguments to the effect that the obvious methods whereby one might imagine distinguishing between identical classical particles are untenable. It is my fond but possibly forlorn hope that those arguments will hasten the ultimate demise of the misconception that identical classical particles are somehow distinguishable. This misconception has created an artificial and entirely unnecessary dichotomy between classical and quantum statistical mechanics, thereby obscuring the conceptual coherence of the whole subject and making it more difficult to understand. Conversely, the fundamental principle that all identical particles are inherently and unconditionally indistinguishable unifies and simplifies classical and quantum statistical mechanics, which no longer require separate consideration and coalesce into a single unified treatment. The Gibbs paradox and its infamous *ad hoc* $1/N!$ correction factor are thereby entirely circumvented, and the resulting formulation yields the further insight that the essential distinction between classical and quantum statistics does not reside in the alleged distinguishability of classical particles, but rather in their statistical independence.

Finally, it is recommended that this book be accessed serially or sequentially rather than randomly or selectively, for only in that manner does the logical structure and coherence of the material fully reveal itself. References specifically cited in the text are numbered and listed at the back of the book in the order in which they appear, but references to basic, familiar, and/or well known formulae and results are sporadic. General references wherein useful discussions of such topics can be found are listed in a supplementary Bibliography which precedes the numbered References.

Chapter 2

FUNDAMENTALS

2.1 Entropy as Uncertainty

The most common and popular description of entropy in everyday language is that it provides a measure of disorder. Other terms used to convey an intuitive feeling for entropy include randomness, disorganization, "mixed-up-ness" (Gibbs), missing information, incomplete knowledge, complexity, chaos, ignorance, and uncertainty. Of course, such descriptors are inherently imprecise and qualitative, and are inadequate to provide a precise quantitative definition, but the vernacular term that we believe best captures the essence of entropy is *uncertainty*. ("Ignorance" would be equally accurate, but seems less suitable due to its negative connotations.)

Thus we shall regard and approach entropy as a quantitative measure of uncertainty. Uncertainty is clearly related to the number of possibilities that exist, and would be expected to increase as that number becomes larger. It also seems natural to define the collective uncertainty of two or more independent or unrelated situations as the sum of their individual uncertainties, so that uncertainty is additive. In the present context, uncertainty pertains to which of its accessible states a particular system occupies. Thus we seek a quantitative measure of uncertainty which is additive over statistically independent systems.

2.2 Systems and States

The concept of the "state" of a system is of fundamental importance in many areas of physics, and lies at the very heart of entropy. As will be seen, evaluating the entropy requires identifying and labeling the states accessible to the system, as well as determining or assigning their probabilities. Thus it is essential at the outset to have a clear conception of what does and does not constitute a "system" and a "state," and to draw a sharp distinction between the two. An intuitive understanding of these concepts is more easily conveyed by simple examples than by abstract general definitions. Consider a single golf ball which is known to be in one of the eighteen holes of a golf course. The system is the ball itself, the states are the holes, and the uncertainty pertains to which hole contains the ball. The simplest systems which manifest nonzero uncertainty are those with only two possible states, such as a ceiling light in a closed closet. The system is the light, its two possible states are "on" or "off," and the associated uncertainty pertains to whether or not the light has been left on. Thus the state of a system is in essence a full and complete description or specification of its current condition, circumstances, configuration, arrangement, etc. The associated uncertainty clearly depends on the relative likelihood of the possible states. Thus the uncertainty as to the state of the aforementioned closet light is relatively small if those who use it usually remember (or usually forget!) to switch it off, but much larger if they are more or less equally likely to remember or forget.

In more general terms, the state of a system is defined by a minimal set of independent parameters or variables whose values are sufficient to completely specify the condition of the system, in the sense that all of its observable and/or measurable properties (including its future time evolution in dynamical problems) are determined by its state. It is essential to note that *by their very nature, the states of a system are inherently distinct and can be regarded as labeled or numbered*; otherwise it would be meaningless to say that a system is in a particular state. The states of a system may be either discrete (e.g., the six faces of a die) or continuous (e.g., the points on a line), or even a combination of the two. It is convenient to label discrete states with the integers $k = 1, 2, \cdots, W$, where W is the total number of states accessible to the system. Continuous states must be labeled by continuous parameters or variables. Bound states are discrete in quantum mechanics, but discrete states are common in classical systems as well (e.g., the aforementioned die). Thus the term "state" and the concept it represents are of very wide and general applicability, and must not be considered synonymous with either "discrete state" or "quantum state."

In contrast to the states they occupy, the objects, elements, or particles of which the system is composed need not, and generally will not, be labeled. Indeed, identical particles (e.g., atoms or molecules of the same type) *cannot* be labeled even in principle (except in a fictitious *gedanken* sense, as will be discussed in due course), for if they were they would no longer be identical. This point of view is of course fully consistent with quantum mechanics, but it is equally valid and applicable in classical mechanics too, as will be discussed in detail in Chapter 8.

In accordance with the principle that nothing is ever simple, it must also be emphasized that the same system may be described in greater or lesser degrees of detail, and its states are implicitly defined relative to and in terms of the level of the description employed. Thus the same system may possess more than one type of state description depending on the context. For present purposes, the relevant example is a macroscopic system composed of a very large number of particles (atoms or molecules). Such systems can be described either in terms of their macroscopic thermodynamic variables or their microscopic molecular variables. The thermodynamic state of a system of N identical particles can be defined by (N, V, E), where V is the volume they occupy and E is the internal energy, because all other thermodynamic variables (e.g., pressure and temperature) can be expressed as definite functions of (N, V, E) by means of the thermodynamic equations of state. In classical mechanics, the microscopic state of such a system is typically defined by specifying the positions and velocities (or momenta) of all the particles, which constitute the minimal information required to uniquely determine the future dynamical time evolution of the system. It is natural and conventional to refer to these two types of states as *macrostates* and *microstates*, respectively. The latter description is obviously much more detailed, and it is clear that the number of microstates consistent with a given macrostate is normally extremely large. The statistical entropy of a system is fundamentally defined in terms of its microstates, while the role played by the macrostates is to impose appropriate constraints on the microstates. We emphasize, however, that while macroscopic thermodynamic systems constitute its most important application, statistical entropy is a more general construct of much wider applicability.

2.3 The Boltzmann-Planck Entropy

The simplest possible manifestation of uncertainty would seem to be that associated with a system which is equally likely to occupy or be found in any one of W possible states. The assumption that such is the case is traditionally referred to as the hypothesis of equal *a priori* probabilities (EAPP). A discussion of the conditions under which the EAPP hypothesis may reasonably be expected, or at least provisionally presumed, to be valid will be deferred to Chapter 3. For the present, we simply proceed to consider how the uncertainty of such a situation might be quantified. The only available parameter is the number of states W, so the uncertainty or entropy S must be a universal function of W; i.e.,

$$S = \sigma(W) \tag{2.1}$$

When $W = 1$ there is only a single state accessible to the system, in which case the system must occupy that state and there is no uncertainty. Thus we require $\sigma(1) = 0$. Moreover, as the number of accessible states increases, the uncertainty as to which state the system occupies likewise increases. It follows that $\sigma(W)$ must be a monotonically increasing function of W which vanishes for $W = 1$.

Now consider two separate noninteracting systems, systems $\mathcal{A}$ and $\mathcal{B}$, each of which is known to occupy one of $W_{\mathcal{A}}$ and $W_{\mathcal{B}}$ equally probable states, respectively. Their corresponding uncertainties are $S_{\mathcal{A}} = \sigma(W_{\mathcal{A}})$ and $S_{\mathcal{B}} = \sigma(W_{\mathcal{B}})$. It is natural to define the total uncertainty of both systems together to be the sum of their separate uncertainties; i.e.,

$$S_{\mathcal{AB}} = S_{\mathcal{A}} + S_{\mathcal{B}} = \sigma(W_{\mathcal{A}}) + \sigma(W_{\mathcal{B}}) \tag{2.2}$$

However, the two systems together can also be regarded as a single combined system $\mathcal{AB}$. Now if system $\mathcal{A}$ is in state $k_{\mathcal{A}}$ and system $\mathcal{B}$ is in state $k_{\mathcal{B}}$, the state of the combined system $\mathcal{AB}$ is defined by the ordered pair $(k_{\mathcal{A}}, k_{\mathcal{B}})$. That is to say, the state space of the combined system is the Cartesian product of the state spaces of the individual systems. The number of accessible states of the combined system is the number of such ordered pairs, which is clearly $W_{\mathcal{AB}} = W_{\mathcal{A}} W_{\mathcal{B}}$. Since the two systems are independent, the probability of the combined state $(k_{\mathcal{A}}, k_{\mathcal{B}})$ is simply $p(k_{\mathcal{A}}, k_{\mathcal{B}}) = (1/W_{\mathcal{A}})(1/W_{\mathcal{B}})$, which is independent of $(k_{\mathcal{A}}, k_{\mathcal{B}})$. The states of the combined system are therefore equally probable, so the associated uncertainty or entropy can also be computed as

$$S_{\mathcal{AB}} = \sigma(W_{\mathcal{AB}}) = \sigma(W_{\mathcal{A}} W_{\mathcal{B}}) \tag{2.3}$$

In order for Eqs. (2.2) and (2.3) to be consistent, the function σ must have the property

$$\sigma(xy) = \sigma(x) + \sigma(y) \tag{2.4}$$

for any two positive integers x and y. It is obvious by inspection that the function

$$\sigma(W) = \sigma_0 \log W \tag{2.5}$$

satisfies Eq. (2.4), where σ_0 is an arbitrary constant and $\log x$ is the natural logarithm of x. (The logarithm to any other base could equally well be used, but this would merely modify the value of σ_0, which is already arbitrary. Thus no generality is lost by the exclusive use of natural logarithms.) Notice that Eq. (2.5) satisfies the previously stated requirement that $\sigma(W)$ be a monotonically increasing function of W which vanishes for $W = 1$.

However, it is not obvious that the functional form of Eq. (2.5) is uniquely determined by Eq. (2.4). In most physical applications W is so enormously large that it can be regarded as a continuous variable for all practical purposes. If the range of the variables x and y is accordingly extended from the positive integers to the positive real numbers, and if the function $\sigma(x)$ is required to be continuous and differentiable, then uniqueness is easily demonstrated by differentiating Eq. (2.4) with respect to x to obtain

$$\sigma'(x) = y\,\sigma'(xy) \tag{2.6}$$

where $\sigma'(x) \equiv d\sigma(x)/dx$. Interchanging x and y we likewise obtain $\sigma'(y) = x\,\sigma'(xy)$, which combines with Eq. (2.6) to imply $x\,d\sigma(x)/dx = y\,d\sigma(y)/dy$. This equates a function of x alone to a function of y alone, both of which must therefore have the same constant value, call it σ_0. Thus $x\,d\sigma(x)/dx = \sigma_0$, which immediately integrates to

$$\sigma(x) = \sigma_0 \log x + C \tag{2.7}$$

where C is a constant of integration. Combining Eqs. (2.4) and (2.7), we obtain $C = 2C$, so that $C = 0$ and Eq. (2.7) reduces to Eq. (2.5). A more general proof of uniqueness which does not assume differentiability or even continuity is given by Balian [18] and Jaynes [19].

The constant σ_0 is arbitrary and can be chosen at will. The choice of σ_0 that ultimately makes S consistent with the conventional thermodynamic entropy is Boltzmann's constant. However, this is a mere historical artifact of the fact that temperature was defined and assigned arbitrary independent units before its significance was fully understood. If temperature is measured in energy units then Boltzmann's constant is unity and entropy becomes dimensionless. We shall adopt this convention by setting $\sigma_0 = 1$. Equations (2.1) and (2.5) then combine to yield

$$S = \log W \tag{2.8}$$

which we shall refer to as the Boltzmann-Planck (BP) entropy of a system with W equally probable states. The problem of evaluating the entropy then reduces to that of counting the states.

As previously remarked, in most cases of physical interest W is extremely large, which has the ironic consequence that W can be in error by a relatively large factor without introducing a significant error into S. If $\widetilde{W}$ is some approximation to the true value of W and $R = \widetilde{W}/W$ is the factor by which $\widetilde{W}$ is in error, then the corresponding approximation to S is

$$\widetilde{S} = \log \widetilde{W} = \log RW = \log R + \log W = S + \log R \tag{2.9}$$

The error in S is then simply $\log R$, which is negligible as long as $|\log R| \ll \log W$. In contrast, the error in W itself is negligible only when $R = \exp(\log R) \cong 1$, which requires $|\log R| \ll 1$. The latter condition on $\log R$ is much more restrictive than the former when $\log W$ is large. This can be concisely expressed by the statement that accuracy of S requires only logarithmic accuracy of W. More generally, it must be kept in mind that logarithmic inequalities, like liberty, demand eternal vigilance, because $a \ll b$ does not imply $\log a \ll \log b$, and $\log a \cong \log b$ does not imply $a \cong b$. For example, if $a = 10^{300}$ and $b = 10^{303}$ then $\log b$ differs from $\log a$ by only 1%, while b differs from a by a factor of 1000. A more pertinent example in the present context is provided by Stirling's approximation $N! \cong \sqrt{2\pi N}(N/e)^N$, which is remarkably accurate even for small values of N. For $N = 10^{23}$, $\log N! \sim 5 \times 10^{24}$ and $\sqrt{2\pi N} \sim 10^{12}$. Omitting the prefactor $\sqrt{2\pi N}$ in the approximation therefore incurs an error of a factor $\sim 10^{12}$ in $N!$ itself, but the corresponding error in $\log N!$ is only $\log \sqrt{2\pi N} \sim 30$ and is therefore utterly negligible.

2.4 The Boltzmann-Gibbs-Shannon Entropy

Consider a cubical die, with its six faces numbered or labeled from $k = 1$ to 6. If the die is true (i.e., perfectly balanced and symmetric) then when it is thrown the probability p_k that it will come to rest with face k upward is independent of k; i.e., $p_k = p$. Those equal probabilities must sum to unity, so $\sum_k p_k = 6p = 1$ and $p_k = p = 1/6$. The number of possible states or outcomes is $W = 6$, so the entropy or uncertainty as to the outcome of a single throw is simply $S = \log 6$. But now suppose that the die is asymmetric and is therefore not true, so that the probabilities p_k are not all equal and some faces are more likely to face upward than others. The numerical values of the probabilities can be determined empirically by repeatedly throwing the biased die a great many times. But once this has been done, we no longer have a basis for assigning a numerical value to the uncertainty of the outcome. The states are no longer equally probable, so Eq. (2.8) no longer applies. It seems intuitively clear that the outcome is now less uncertain than it was before, but to what degree? The uncertainty must clearly be some function of the probabilities p_k, since there are no other variables upon which it might depend, but there is no apparent way of inferring the form of that function from Eq. (2.8).

More generally, if our entire knowledge of a system consists of a list of its accessible states k and their unequal probabilities p_k $(k = 1, \cdots, W)$, we have as yet no prescription whereby we might quantify the associated uncertainty. In short, the BP entropy requires generalization to deal with unequally probable states. Just as the entropy is a universal function of W for equally probable states, so too must its generalization be a universal function of the probabilities p_k for unequally probable states. Since that function is universal, it is uniquely determined once it has been defined for any particular situation in which the values of W and p_k are arbitrary, subject to the normalization condition $\sum_k p_k = 1$. To that end we shall envision each unequally probable state k as a "black box" which contains W_k unobservable substates, all of which are equally probable. The total number of such substates accessible to the system is then $W_t \equiv \sum_k W_k$, and the probability that the system occupies any particular one of them is $1/W_t$. The probability p_k that the system is in state k is simply the probability that it occupies a substate within state k, which is clearly $p_k = W_k(1/W_t) = W_k/W_t$. According to Eq. (2.8), the total entropy or uncertainty as to which of the W_t equally probable substates the system occupies is simply $S_t = \log W_t$. The entropy or uncertainty as to which substate is occupied by a system known to be in state k is likewise simply $S_k = \log W_k$. Our objective is to define an entropy S which represents only the uncertainty as to which state k the system resides in, regardless of which internal substate thereof it may occupy.

Since we require uncertainties and entropies to be additive, it is natural to regard the uncertainty as to which substate the system occupies as the sum of the uncertainty as to which state k it is in and the further uncertainty as to which substate it occupies within state k. Stated in terms of entropies, this relation becomes $S_t = S + S_k$, which would seem to imply that S can simply be evaluated as $S = S_t - S_k$. But this cannot be done, because the state k the system occupies, and consequently its "internal" entropy S_k, is unknown *a priori*. Thus we find ourselves in a "Catch-22" type of situation, in which the state k must be known with certainty in order to evaluate its uncertainty! This dilemma cannot be resolved without adopting a further postulate, of which the obvious choice is to replace the unknown value of S_k by its average over all states k, weighted by their respective probabilities p_k. Thus we set $S = S_t - \sum_k p_k S_k$, which immediately reduces to

$$S = \log W_t - \sum_k (W_k/W_t) \log W_k = -\sum_k p_k \log p_k \qquad (2.10)$$

Equation (2.10) expresses S as a function of the probabilities p_k alone, and thereby defines the desired universal function of the p_k which uniquely determines the entropy of any system with unequally probable states k, regardless of whether or not those states really are composed of equally probable substates. Equation (2.10) is the usual familiar form for the entropy of such systems, and will henceforth be referred to as the Boltzmann-Gibbs-Shannon (BGS) entropy. It is inherently nonnegative because $p_k \leq 1$, so that $p_k \log p_k \leq 0$. Note that the BGS entropy properly reduces to the BP entropy of Eq. (2.8) when the states are equally probable; i.e., when $p_k = 1/W$ for all k. It is also easy to verify that like the BP entropy, the BGS entropy is additive over statistically independent systems. Since $x \log x \to 0$ as $x \to 0$, highly unlikely states make a negligible contribution to S and can therefore be omitted from consideration, just as one would intuitively expect.

Equation (2.10) shows that the entropy S is the statistical average of the quantity $-\log p_k$ over all states, and that each state k contributes an amount $-p_k \log p_k$ to S. However, neither of those observations should be misinterpreted as implying that each state k possesses an entropy of its own, which is decidedly and emphatically *not* the case! If this were true then entropy would be a state variable with a definite value for a system known to be in a particular state k. But if the system were known to be in a particular state there would be no uncertainty as to what state it occupies, so that its entropy would be zero. Thus it is entirely erroneous to think of each particular state as possessing its own intrinsic entropy. Entropy is rather a collective property of an entire group of states, which is determined by their probability distribution according to Eq. (2.10).

In summary, the evaluation of entropy requires one to (a) identify all the states k accessible to the system, and (b) determine or assume their probabilities p_k. Step (a) is normally based on an assumed model or physical picture of the internal structure of the system. If the states are equally probable, the evaluation reduces to counting the states (which however is not always easy) and using Eq. (2.8); otherwise Eq. (2.10) must be used.

2.5 Relative Entropies

Since the BGS entropy is fundamentally defined in terms of the BP entropy, it is instructive to examine the difference between the two. To this end, we observe that Eq. (2.10) can be rewritten as

$$S = \log W - \Delta S(p_k|1/W) \tag{2.11}$$

where

$$\Delta S(p_k|q_k) \equiv \sum_k p_k \log(p_k/q_k) \tag{2.12}$$

is the *relative entropy* of the probability distribution p_k with respect to another normalized probability distribution q_k.[†] (The quantity ΔS also has important applications in statistics and information theory, where it is generally referred to by the imposing appellation of the "Kullback-Leibler divergence.") Note that $S(p_k|q_k)$ is asymmetrical in p_k and q_k. Thus $\Delta S(p_k|1/W)$ is the relative entropy of a system with unequally probable states with respect to an otherwise similar system with equally probable states. The easily verified inequality $\log x \leq x - 1$ combines with Eq. (2.12) to imply that $\Delta S(p_k|q_k) \geq 0$ and vanishes only when $q_k = p_k$. It then follows from Eqs. (2.10) and (2.11) that

$$S = -\sum_k p_k \log p_k \leq \log W \tag{2.13}$$

[†]Here and henceforth the vertical bar | in the notation $f(x|y)$ is simply a variable separator which serves to focus attention on the functional dependence of the quantity f upon the variables x for given values of the variables y. Thus the latter can be regarded as parameters which are temporarily held fixed. This convention is a natural extension of the usual notation for conditional probabilities [19].

so that S is a maximum when $p_k = 1/W$, in which case it properly reduces to $\log W$. Thus the BGS entropy is maximized when the states are equally probable. Equation (2.13) provides our first glimpse of the principle of maximum entropy (PME), which is of the highest importance and will be discussed in Chapter 4. Equation (2.11) shows that the BGS entropy of the probability distribution p_k can be interpreted as the BP entropy of the uniform distribution $1/W$ minus the relative entropy of p_k with respect to $1/W$. Equation (2.13) then confirms our previous intuitive assertion that the uncertainty is largest when all possibilities are equally likely.

In a set of unequally probable states k, the values p_k of their probabilities need not, and in general will not, all be distinct. When such is the case, it is often useful to partition the states k into subsets of equally probable states by grouping together all states which share the same probabilities. Such groups or subsets will be labeled by an index ν, and are presumed to be mutually exclusive and exhaustive. Thus some of the resulting groups may, and in general will, contain only a single state. The number of equally probable states in group ν is denoted by W_ν. The common probability of each of those states is denoted by q_ν, so that $p_k = q_\nu$ for $k \in \nu$. The total number of states k in all the groups is simply $W = \sum_k (1) = \sum_\nu W_\nu$. The relation between such groups and the individual states is analogous to that between energy levels and degenerate states in quantum mechanics. The quantities W_ν are analogous to degeneracy factors. The probability that the state of the system lies in group ν is clearly $p_\nu = W_\nu\, q_\nu$, which therefore defines the probability distribution over the groups. The BGS entropy of Eq. (2.10) can now be reexpressed as

$$\begin{aligned} S &= -\sum_k p_k \log p_k = -\sum_\nu \sum_{k \in \nu} q_\nu \log q_\nu \\ &= -\sum_\nu W_\nu\, q_\nu \log q_\nu = -\sum_\nu p_\nu \log \frac{p_\nu}{W_\nu} \end{aligned} \tag{2.14}$$

which can be rewritten in the form

$$S = \log W - \Delta S(p_\nu | W_\nu / W) \tag{2.15}$$

Since $\Delta S(p_\nu | W_\nu / W) \geq 0$, the probability distribution p_ν which maximizes S and thereby the uncertainty is $p_\nu = W_\nu / W$, which however is not uniform in ν unless W_ν is independent of ν. However, the distribution $p_\nu = W_\nu / W$ implies that $q_\nu = 1/W$ *is* independent of ν, which in turn implies that the individual states k are equally probable, so that their corresponding probability distribution is again the uniform distribution $p_k = 1/W$.

2.6 Continuous States and Probability Densities

In many cases the states available to a system of interest are continuous rather than discrete; i.e., they are defined and labeled by a continuous parameter x or parameters $\mathbf{x} = (x_1, x_2, ...)$ rather than a discrete index k. Perhaps the simplest example is that of a point particle whose accessible states are the geometrical points in the interval $0 \leq x \leq 1$ on the real line. This situation is easily visualized, but it cannot be described within the existing framework described above because the number of states x is infinite. The probability of any particular state (or value of x) is correspondingly zero, so that both the BP and BGS entropies diverge.

Of course, the proper way to describe probabilities in such a situation is in terms of a probability density function $p(\mathbf{x})$, where $p(\mathbf{x})\, d\mathbf{x}$ is the probability that the state of the system lies in the interval $(\mathbf{x}, \mathbf{x} + d\mathbf{x})$. Note that in spite of the notation and in contrast to p_k, $p(\mathbf{x})$ is not itself a probability, and this must be kept in mind. In order to describe continuous state spaces and probability distributions, it is therefore necessary to formulate suitable generalizations of the BP and BGS entropies which are sensible and well defined in that context.

The BP entropy is a special case of the BGS entropy, so it suffices to consider the latter.

The obvious generalization of Eq. (2.10) to continuous states would naively seem to be

$$S = - \int d\mathbf{x}\, p(\mathbf{x}) \log p(\mathbf{x}) \tag{2.16}$$

which is evidently finite. Unfortunately, this superficially plausible expression is untenable for several related reasons: (a) The BGS entropy of Eq. (2.10) is specifically defined in terms of probabilities, whereas $p(\mathbf{x})$ is not a probability. (b) The variable $\mathbf{x}$ is not in general dimensionless so neither is $p(\mathbf{x})$, but the argument of a logarithm is required to be dimensionless. (c) A proper general expression for the entropy in terms of $p(\mathbf{x})$ must reduce to Eq. (2.10) in the special case of discrete states; i.e., $p(\mathbf{x}) = \sum_k p_k \delta(\mathbf{x} - \mathbf{x}_k)$, where $\delta(\mathbf{x})$ is the Dirac delta function. However, logarithms involving delta functions are not well defined. (d) Equation (2.16) is not covariant with respect to an invertible nonlinear transformation from the variable $\mathbf{x}$ to an alternative equivalent state variable $\mathbf{y} = \mathbf{f}(\mathbf{x})$ [19–21]. Since $\mathbf{x}$ uniquely determines $\mathbf{y}$ and vice versa, the state of the system can equally well be specified by the value of $\mathbf{y}$, so the entropy should equally well be given by $S = - \int d\mathbf{y}\;\; p(\mathbf{y}) \log p(\mathbf{y})$. However, this differs from Eq. (2.16) by an amount $\int d\mathbf{x}\; p(\mathbf{x}) \log J(\mathbf{x})$, where $J \equiv \|\partial \mathbf{y}/\partial \mathbf{x}\|$ is the Jacobian determinant of the transformation. The entropy cannot be allowed to depend on the arbitrary choice of which nonunique continuous variable is used to label the states of the system. Thus Eq. (2.16) is unacceptable as a general definition of the entropy of a system with a continuous state space.

These difficulties evidently originate in the fact that the BGS entropy is defined in terms of the BP entropy, which is meaningful only when the number of states is finite and diverges when it is not. In order to circumvent this dilemma, we shall adopt the intuitively sensible position that it is not operationally meaningful to distinguish between states which are so close together that they are not significantly or sensibly different. Viewed from this perspective, the *de facto* state of a system is determined and specified for all practical purposes if the system is known to be in the immediate vicinity or neighborhood of a given state point $\mathbf{x}$. This suggests that states within a very small neighborhood of $\mathbf{x}$ should be grouped together and regarded as being effectively a single *de facto* state. The obvious way to implement this idea is to discretize state space by dividing it into very small subregions $\Delta\mathbf{x}_k$, each of which is regarded as a discrete state. However, it is simpler and more elegant to simply introduce a smooth continuous density of states $\rho(\mathbf{x})$ defined in such a way that the finite number of such *de facto* states in a small but finite interval $(\mathbf{x}, \mathbf{x}+\Delta\mathbf{x})$ is closely approximated by $\rho(\mathbf{x})\,\Delta\mathbf{x}$ for sufficiently small $\Delta\mathbf{x}$. Since they are virtually identical, each of those states must have very nearly the same probability $q(\mathbf{x})$, which is also presumed to be a smooth continuous function. Note that in contrast to $p(\mathbf{x})$, $q(\mathbf{x})$ actually is a probability and is consequently dimensionless for any choice of the state variables $\mathbf{x}$. The probability that the state of the system lies in the interval $(\mathbf{x}, \mathbf{x}+\Delta\mathbf{x})$ is then $\rho(\mathbf{x})\,q(\mathbf{x})\,\Delta\mathbf{x}$, so that the probability density is simply

$$p(\mathbf{x}) \;=\; \rho(\mathbf{x})\,q(\mathbf{x}) \tag{2.17}$$

and conversely $q(\mathbf{x}) = p(\mathbf{x})/\rho(\mathbf{x})$. The functions $\rho(\mathbf{x})$ and $q(\mathbf{x})$ must therefore satisfy the normalization condition

$$\int d\mathbf{x}\;\rho(\mathbf{x})\,q(\mathbf{x}) \;=\; 1 \tag{2.18}$$

We can now evaluate the BGS entropy from Eq. (2.10) by means of the obvious replacements $\sum_k \to \int d\mathbf{x}\;\rho(\mathbf{x})$ and $p_k \to q(\mathbf{x})$ therein, with the result

$$\begin{aligned} S &= -\int d\mathbf{x}\;\rho(\mathbf{x})\;q(\mathbf{x})\log q(\mathbf{x}) \\ &= -\int d\mathbf{x}\;p(\mathbf{x})\log\frac{p(\mathbf{x})}{\rho(\mathbf{x})} \end{aligned} \tag{2.19}$$

Equation (2.19) is the proper generalization of the BGS entropy to systems with continuous states.

Discrete states can be regarded as a special case of continuous states, in which $\rho(\mathbf{x})$ assumes the form

$$\rho(\mathbf{x}) = \sum_k \delta(\mathbf{x} - \mathbf{x}_k) \tag{2.20}$$

Equation (2.17) then becomes

$$p(\mathbf{x}) = \sum_k q(\mathbf{x}_k)\,\delta(\mathbf{x} - \mathbf{x}_k) = \sum_k p_k\,\delta(\mathbf{x} - \mathbf{x}_k) \tag{2.21}$$

where $p_k = q(\mathbf{x}_k)$ is simply the probability of the single discrete state $\mathbf{x}_k$. The first equality in Eq. (2.19) similarly reduces to

$$S = -\sum_k q(\mathbf{x}_k) \log q(\mathbf{x}_k) = -\sum_k p_k \log p_k \tag{2.22}$$

in agreement with Eq. (2.10).

The function $\rho(\mathbf{x})$ has variously been referred to as a "measure function" or the distribution of "complete ignorance" [19–21]. We prefer the terminology "density of states," which is precisely what $\rho(\mathbf{x})$ represents. The number of states in a region $\mathbb{R}$ is clearly $\int_{\mathbb{R}} d\mathbf{x}\ \rho(\mathbf{x})$, so the total number of states in the entire state space is simply given by

$$W = \int d\mathbf{x}\ \rho(\mathbf{x}) \tag{2.23}$$

Thus $\rho(\mathbf{x})$ is not itself a normalized probability density, but the normalized probability density corresponding to $\rho(\mathbf{x})$ is simply

$$p_0(\mathbf{x}) \equiv (1/W)\,\rho(\mathbf{x}) \tag{2.24}$$

If all states are equally probable then $q(\mathbf{x})$ is independent of $\mathbf{x}$ and has the value $1/W$, whereupon Eq. (2.17) reduces to $p(\mathbf{x}) = p_0(\mathbf{x})$

and Eq. (2.19) reduces to the BP entropy $S = \log W$. Thus $p_0(\mathbf{x})$ is the probability density for equally probable continuous states, and is therefore the continuous analog of the uniform discrete probability distribution $p_k = 1/W$.

In contrast to the untenable Eq. (2.16), Eq. (2.19) is properly covariant with respect to nonlinear transformations of the state variables $\mathbf{x}$, because both $p(\mathbf{x})$ and $\rho(\mathbf{x})$ transform in the same way and the Jacobian J cancels out. However, Eq. (2.19) depends on $\rho(\mathbf{x})$ as well as $p(\mathbf{x})$, and hence is ambiguous until $\rho(\mathbf{x})$ has somehow been determined. Some authors suggest or imply that this ambiguity is peculiar to continuous states and does not occur for discrete states, but we beg to differ; it is entirely analogous to the ambiguity which would arise in the discrete case if Eq. (2.14) were used to express the entropy in terms of the group probabilities p_ν rather than the state probabilities p_k and the quantities W_ν were unknown. Indeed, Eq. (2.19) is the obvious continuous analog of Eq. (2.14), which has the same mathematical structure. Equation (2.19) can be rewritten in the equivalent form

$$S = \log W - \int d\mathbf{x}\, p(\mathbf{x}) \log \frac{p(\mathbf{x})}{p_0(\mathbf{x})} \tag{2.25}$$

The inequality $\log x \leq x - 1$ combines with Eqs. (2.19) and (2.25) to imply that

$$S = -\int d\mathbf{x}\, p(\mathbf{x}) \log \frac{p(\mathbf{x})}{\rho(\mathbf{x})} \leq \log W \tag{2.26}$$

so that S attains its maximum value of $\log W$ when $p(\mathbf{x}) = p_0(\mathbf{x})$. Equation (2.26) is the continuous analog of Eq. (2.13) and the associated maximum principle for the discrete BGS entropy.

The functional form of the probability density $p(\mathbf{x})$, its value at a given state point, and its dimensions all depend on the choice of the state variables $\mathbf{x}$. In contrast, $q(\mathbf{x}) = p(\mathbf{x})/\rho(\mathbf{x})$ is dimensionless and fully covariant; it has the same numerical value at a given point in state space regardless of the choice of the state variables. Since the number of states in the interval $(\mathbf{x}, \mathbf{x} + d\mathbf{x})$ is simply $dW = \rho(\mathbf{x})\, d\mathbf{x}$, Eq. (2.19) can be rewritten in the manifestly covariant form

$$S = -\int dW\, q \log q \tag{2.27}$$

When the states are equally probable, $q(\mathbf{x}) = 1/W$ and Eq. (2.27) immediately reduces to $S = -(1/W)\log(1/W)\int dW = \log W$ as before. Since $d\mathbf{x}$ is the Cartesian volume element in $\mathbf{x}$-space, dW can also be interpreted as a covariant volume element in state space. It then follows that equal volumes in state space contain equal numbers of states [22].

Unfortunately, there is no general theory or procedure for determining $\rho(\mathbf{x})$ for any particular choice of the state variables $\mathbf{x}$. In the absence of arguments or evidence to the contrary, one might at first be tempted to simply presume that the states are distributed uniformly in $\mathbf{x}$-space, so that $\rho(\mathbf{x})$ is a constant ρ_x independent of $\mathbf{x}$. (The subscript x on ρ_x is not a variable or argument but merely a label which serves to identify the particular set of variables $\mathbf{x}$.) However, this would again lack covariance with respect to a nonlinear transformation $\mathbf{y} = \mathbf{f}(\mathbf{x})$, which transforms a uniform density of states $\rho(\mathbf{x}) = \rho_x$ in the variables $\mathbf{x}$ into the nonuniform density of states $\rho(\mathbf{y}) = \rho_x \|\partial\mathbf{x}/\partial\mathbf{y}\|$ in the variable $\mathbf{y}$. Thus a uniform density of states in $\mathbf{x}$ is sensible only if it is based on arguments or considerations which specifically suggest or identify a particular preferred choice of the variables $\mathbf{x}$. When such is the case, the variables $\mathbf{x}$ will be referred to as a "natural representation" of the state of the system [23].

The functional form of $\rho(\mathbf{x})$ can sometimes be inferred from group-invariance arguments [19–21], the basic idea of which is that the maximum uncertainty or complete ignorance represented by $\rho(\mathbf{x})$ should be invariant with respect to transformations which essentially describe changes in viewpoint or perspective. Once $\rho(\mathbf{x})$ has thereby been determined for any particular choice of the variables $\mathbf{x}$, one can proceed to simplify the description by transforming to a natural representation in which $\rho(\mathbf{x})$ is constant. Group invariance defines a certain type of symmetry, and its use to determine $\rho(\mathbf{x})$ represents a generalization of the observation that equally probable discrete states are typically a consequence of symmetry (e.g., the six faces of a perfectly symmetrical cubical die). Indeed, symmetry would seem to provide the only objective basis from which equal *a priori* probabilities (EAPP) can rigorously be deduced. Unfortunately, complete symmetry between accessible states rarely obtains in systems of practical interest, a circumstance which normally relegates EAPP to the status of a hypothesis rather than a theorem, as discussed in the next chapter.

2.7 Systems with an Infinite Number of States

In many systems of practical interest, the states are in principle infinite in number even when they are discrete. For example, a quantum harmonic oscillator has an infinite number of states $k = 1, 2, \cdots$ with corresponding energies $E_k = (2k - 1)E_1$. More mundane examples of countably infinite states include the steps on a semi-infinite ladder, the squares of an infinite checkerboard, or simply the positive integers $k = 1, 2, \cdots$ themselves. When the number of states is infinite the BP entropy of Eq. (2.8) diverges. This merely reflects the obvious fact that an infinite number of states cannot all have the same nonzero probability. Their probabilities must therefore differ, so the entropy of the system must be evaluated as the BGS entropy of Eq. (2.10), the summation in which then becomes an infinite series. The BGS entropy S is then finite only when the probabilities p_k have values for which that series converges. When such is the case, the resulting convergent series can be approximated to arbitrary accuracy by a finite number of terms. The remaining infinity of states that contribute negligibly to S are not strictly inaccessible, but they are effectively inaccessible since the sum of their probabilities is negligible by virtue of the comparison test. Entirely similar considerations apply when the states are continuous, in which case the probability density $p(\mathbf{x})$ must approach zero sufficiently rapidly as $|\mathbf{x}| \to \infty$ that the integral in Eq. (2.19) is finite.

The simplest way to ensure the convergence of the series in Eq. (2.10) is to impose conditions or restrictions which render all but a finite number W of the states k strictly inaccessible, so that the probabilities p_k of all the other states vanish. In the special case when those W accessible states are equally likely, their probabilities are simply $p_k = 1/W$ and Eq. (2.10) reduces to the BP entropy $S = \log W$. In the continuous case, the number of accessible states can be rendered finite by restricting the system to a finite region $\mathbb{R}$ of the state space $\mathbf{x}$. The resulting finite number of states is $W = \int_{\mathbb{R}} d\mathbf{x}\, \rho(\mathbf{x})$, and if those states are equally probable then $p(\mathbf{x}) = p_0(\mathbf{x})$, whereupon Eq. (2.25) again reduces to the BP entropy $S = \log W$. It should be noted that in both the discrete and continuous cases, such accessibility restrictions may be expressed in terms of certain parameters, in which case it must be remembered that the entropy will inherit and retain a functional dependence upon those same parameters.

To recapitulate: if the states of a system are in principle infinite in number, then the values of their probabilities p_k must be such that all but a finite subset of those states become either strictly or effectively inaccessible. Otherwise the problem is ill-posed and improperly formulated, with the pathological consequence that the entropy diverges.

Chapter 3

THE HYPOTHESIS OF EQUAL A PRIORI PROBABILITIES

At this point the task of defining the entropy has been completed, but its full significance has yet to be explored, and the task of actually evaluating it has barely begun. Equation (2.10) expresses the BGS entropy S of a system with discrete states as a function of their probabilities p_k, so S cannot be evaluated until values have been assigned to those probabilities. This evaluation is greatly simplified in the case when all the probabilities are known or presumed to be equal; i.e., when $p_k = 1/W$. The BGS entropy then reduces to the BP entropy of Eq. (2.8), which can be evaluated simply by counting the states. Similarly Eq. (2.19) expresses the BGS entropy of a system with continuous states as a functional of the probability density $p(\mathbf{x})$, which likewise reduces to the BP entropy of Eq. (2.8) when the continuous states are equally probable; i.e., when $p(\mathbf{x}) = p_0(\mathbf{x}) = (1/W)\,\rho(\mathbf{x})$, with W given by Eq. (2.23). In either case, the assumption that the discrete or continuous states of a particular system are equally probable (perhaps subject to certain conditions or restrictions) is traditionally referred to as the hypothesis of equal *a priori* probabilities (EAPP). This hypothesis clearly effects an enormous simplification when W is large. It also implies that the BGS entropy of Eq. (2.10) or (2.19) has its maximum possible value, as discussed in Chapter 2.

The question then arises as to the conditions under which one might reasonably expect or presume the EAPP hypothesis to be valid. Most systems of practical interest are not amenable to a definitive theoretical analysis, so one is usually required to address this question by means of plausibility arguments. Such arguments will be discussed below, and typically provide a rationale for adopting or invoking the EAPP hypothesis on a provisional basis. Its validity can then be assessed *a posteriori* by examining the accuracy of its predictions. Its most conspicuous and best known success is in its role as the very cornerstone of equilibrium statistical mechanics, where its validity is restricted to systems with constant energy for reasons to be elucidated in due course.

3.1 Discrete States

The validity of the EAPP hypothesis is rigorously assured only for systems whose states are equally probable by virtue of being completely symmetrical. Familiar examples of such states include the two sides of a fair coin, or the six faces of a perfectly symmetrical cubical die. It seems intuitively clear that such states must be equally probable, and this expectation is borne out by experience. However, the logical basis for this expectation is a bit subtle and requires some discussion. It is essential at the outset to reemphasize the superficially trivial but fundamental point that the states must somehow be labeled, either explicitly or implicitly, for otherwise it would not be possible to specify which state the system occupies. It may be helpful to envision the system as some object and the states as a collection of boxes, one of which contains the object. The states can then be distinguished by writing or affixing the labels $k = 1, 2, \cdots$ on the sides of the boxes. Note that the labels themselves are not intrinsic properties of the states; any other labels would serve equally well, such as $k = a, b, c, \cdots$ or $k =$ John, Paul, George, $\cdots$. Indeed, the labels are in fact arbitrary and can therefore be permuted, reassigned, redefined, or changed at will, subject only to the requirement that each state must be assigned a unique label which unambiguously identifies it. Once this has been done the states can obviously be distinguished by their labels if nothing else, and therefore cannot be regarded as indistinguishable even if they are similar in all other respects.

However, it should be noted that this simple picture may be obscured by the fact that the states of most systems actually *do* possess intrinsic distinguishing features, in which case it is often convenient to press such features into service as labels, or more precisely to define the labels in terms of them. For example, if there are three boxes which are indelibly colored red (R), green (G), and blue (B), then it is only natural and convenient to simply define their labels as their colors, so that $k = \mathrm{R, G, B}$. The colors then play a dual role as both labels and intrinsic features of the states, but it is important to appreciate that the two are conceptually distinct. This distinction may seem pedantic, but it is essential to a clear understanding of symmetry. It should also be noted that two or more states may share the same intrinsic feature (e.g, there may be two blue boxes), in which case that feature alone clearly cannot serve as a unique label.

Symmetry between states can now be defined as follows: *the states of a system are completely symmetric if their only distinguishing features are their labels*. This definition implies that symmetric states could no longer be distinguished if their labels were removed or erased, in which case it would no longer be possible to determine which state the system is in by observing or inspecting it. The system would then present the same appearance regardless of which state it occupies, and if it were observed a second time it would be impossible to determine whether or not it occupied the same state on both occasions. For example, if one were to erase the dots on the faces of a perfectly symmetrical cubical die and throw it onto a tabletop, the result would look the same regardless of which face is upward. Moreover, if one were to leave the room and return, there would be no way to determine if the upper face of the die (i.e., the state of the system) does or does not differ from what it was before. In contrast, the unlabeled faces of a biased or loaded die sans dots could still be distinguished by experimentally determining the location of its center of mass relative to each face.

Equipped with the above operational definition of symmetric states, we can now proceed to explore its implications for their probabilities. For this purpose, it is essential to clearly understand the further fundamental point that in spite of the notation, *unequal probabilities p_k cannot depend explicitly on the labels k*. The reason is that the probabilities p_k are intrinsic features or properties of the states, whereas the labels k are arbitrary and consequently bear no relation whatever to any intrinsic properties of the states to which they are affixed. Thus, for example, the probabilities for the aforementioned loaded die are determined by its center of mass, and not by what may or may not be inscribed on its faces. This observation implies that any dependence of p_k upon k must be *implicit* in nature, and the only mechanism whereby such a dependence can occur is via a functional dependence of p_k on some other intrinsic property or properties of the states k. Such features or properties are variously referred to as *state functions, state variables*, or *observables*. A generic set of such variables is collectively denoted by $\mathbf{A} \equiv (A_1, A_2, \cdots)$, and their values in state k are denoted by $\mathbf{A}_k$. If no such properties exist, then the probabilities p_k cannot depend on k at all and must consequently be equal. But states with no intrinsic properties could not be distinguished if their labels were removed, and are therefore symmetric by definition. It follows that symmetric states are equally probable, and conversely that unequal probabilities p_k imply asymmetrical states which possess at least one intrinsic observable property A_k. This argument rigorously validates the EAPP hypothesis for systems with fully symmetric states. Unfortunately, such systems are rarely encountered except in games of chance, so the direct applicability of this result is rather limited. However, the reasoning employed in the analysis can be adapted to obtain valuable insight into the validity of the EAPP hypothesis in more general systems, as we now proceed to discuss.

The simplest situations in which unequal probabilities can occur are those in which the states possess only a single intrinsic property A_k. We then have $p_k = \hat{p}(A_k)$, where the nonnegative function $\hat{p}(x)$ depends only on the single argument x and is independent of k. Apart from a multiplicative constant whose value is determined by the normalization condition $\sum_k \hat{p}(A_k) = 1$, the form of the function $\hat{p}(x)$ is indeterminate in the general theory, but this merely reflects the fact that it depends on the nature of the system and states, whether and how the system interacts with other systems or its surroundings, and possibly on its preparation, fabrication, and/or history. Suppose, for example, that the aforementioned boxes are otherwise identical cubes of different sizes. The sole intrinsic property of box k is then its edge length L_k (or an arbitrary function thereof). Depending on the procedure used to place the system into a box, its probability $p_k = \hat{p}(L_k)$ of occupying box k might conceivably be proportional to the volume L_k^3, or the face area L_k^2, or some other function of L_k.

We now observe that regardless of the form of the function $\hat{p}(x)$, the relation $p_k = \hat{p}(A_k)$ implies that all states k having the same value of A_k are equally probable. It follows that if the system is restricted to the subset of states for which $A_k = a$, where a denotes a particular specified value of the variable A_k, then all states in that subset are equally probable, so that the EAPP hypothesis is conditionally valid for that subset of states. This can again be interpreted as a consequence of symmetry, because the states k for which $A_k = a$ cannot be distinguished by their values of A_k, which implies that their only distinguishing features are their labels.

The preceding argument is conceptually sound, but in its present form it suffers from an inconvenient technical flaw, which however can be circumvented by tolerating a certain degree of imprecision. The problem is that the states k and the values of A_k are discrete, whereas the variable a is continuous. The values of a which coincide with any of the discrete values A_k therefore constitute a set of measure zero in a-space, so the subset of states for which $A_k = a$ is actually empty for almost all values of a. This defect can be remedied by introducing a small but finite tolerance into the restriction $A_k = a$ in order to encompass states k for which A_k lies in the immediate neighborhood or vicinity of the specified value a. We therefore replace the restriction $A_k = a$ by the more tolerant restriction

$$A_k \cong a \tag{3.1}$$

in which it is understood that the approximate equality sign $\cong$ is to be interpreted as a shorthand notation for an appropriate definition of the neighborhood of a; e.g., $|A_k - a| \leq \Delta a$. The number of states k in the subset defined by Eq. (3.1) is denoted by $W(a)$. The tolerance associated with and implied by Eq. (3.1) is presumed to be small enough that A_k differs only slightly from a, and yet large enough that $W(a) \gg 1$. The present formulation is sensible and useful only when both of those conditions can be simultaneously satisfied. Of course, the probabilities $p_k = \hat{p}(A_k)$ of the states in the subset defined by Eq. (3.1) will no longer be precisely equal, but if the function $\hat{p}(x)$ is continuous then the approximation $p_k = \hat{p}(A_k) \cong \hat{p}(a)$ will nevertheless be accurate to within a tolerance proportional to $d\hat{p}(a)/da$. The probabilities p_k are then approximately equal, so that the EAPP hypothesis is approximately valid for a system confined to the subset of states defined by the specified value of a. The entropy of such a system is then approximated by the BP entropy $S(a) = \log W(a)$. A more precise and formal treatment of tolerant restrictions such as Eq. (3.1) will be developed and discussed in the next chapter.

In systems of practical importance the states typically possess more than one intrinsic property. The preceding development is easily generalized to accommodate such systems by means of the replacements $A_k \to \mathbf{A}_k$ and $a \to \mathbf{a} \equiv (a_1, a_2, \cdots)$. The probabilities of the individual states k are then given by $p_k = \hat{p}(\mathbf{A}_k)$, whereupon the restrictions

$$\mathbf{A}_k \cong \mathbf{a} \tag{3.2}$$

then imply the approximate validity of the EAPP hypothesis for such systems. Note, however, that any intrinsic properties the states may possess upon which the probabilities p_k do not actually depend (i.e., of which the function $\hat{p}(\mathbf{a})$ is independent) are irrelevant for present purposes and can therefore be removed or omitted from the set **A** at the outset. This point is important, because such properties are often much more numerous than those upon which the probabilities fundamentally do depend, and it would be both nonsensical and unmanageable to include them in the restrictions of Eq. (3.2). It follows that when the states have multiple intrinsic properties, as they ordinarily do, the EAPP hypothesis is contingent on a decision as to which properties are relevant or irrelevant in the sense that the probabilities p_k are or are not considered likely to depend upon them. This question is rarely susceptible to rigorous analysis, so the decision of which properties are relevant and should be included in the set **A** is usually based on plausibility arguments or intuition. This is the point at which EAPP becomes a hypothesis. As such, its validity must be assessed by comparing its predictions with observations or experiments. Any discrepancies encountered in such comparisons should not simply be regarded as mere failures, but rather as useful information which should be subjected to further scrutiny. In particular, they suggest the possibile existence of an additional intrinsic property of the states which was initially overlooked or presumed to be irrelevant but in reality affects the probabilities p_k and should therefore be included in the set **A**.

In most physical systems, including the many-particle systems of primary interest in statistical mechanics, by far the most important intrinsic property of the states k is their energy E_k. The fundamental importance of the energy in this connection is primarily due to its status as a conserved variable, which implies that the energy of an isolated system is constant. Of course, there are other conserved variables as well, notably mass and momentum, but mass is trivially conserved in a closed system composed of a fixed number of particles, and momentum is trivially conserved if the entire system is at rest.

Thus, as Schrödinger [24] has observed, the special status of the energy derives from the fact that it is in general the only simple nontrivial conserved quantity in the system. In contrast, the other intrinsic properties of the states do not ordinarily possess a comparable physical significance. It is therefore reasonable to suppose that the only intrinsic property of the states upon which their probabilities can depend is their energy, so that $p_k = \hat{p}(E_k)$. The fact that the energy of an isolated system is constant then implies that the EAPP hypothesis is valid for isolated systems. Conversely, a system which is not isolated can exchange energy with its surroundings and thereby occupy states with different energies, so that its states can no longer be presumed equally probable.

Thus we see that *in physical systems for which energy is a conserved state variable, the validity of the EAPP hypothesis is plausible only for isolated systems with fixed total energy*. In particular, the EAPP hypothesis serves in precisely that capacity as the cornerstone of the equilibrium statistical mechanics of thermodynamic systems, as will be discussed in due course. Of course, it is necessary to allow for a small tolerance in the specified value of energy, since Eq. (3.1) now becomes $E_k \cong E$. The small uncertainty in energy has the physical interpretation that no physical system is ever perfectly isolated, and even if it were the value of its energy could not be controlled or measured to infinite precision. The EAPP hypothesis then implies that the entropy of the system is approximately given by $S(E) = \log W(E)$, where $W(E)$ is the number of states for which $E_k \cong E$.

3.2 Continuous States

The concepts involved in formulating the EAPP hypothesis for systems with continuous states are closely analogous to those already discussed above for discrete states, but they present some additional nontrivial complications. Those complications originate in the fact that in contrast to discrete states, equally probable continuous states are defined in terms of the density of states $\rho(\mathbf{x})$, which is not known *a priori*. As discussed in Sect. 2.6, the probability density for equally probable continuous states is given by $p_0(\mathbf{x}) = (1/W)\,\rho(\mathbf{x})$. The EAPP hypothesis for continuous states therefore takes the form $p(\mathbf{x}) = p_0(\mathbf{x})$ for $\mathbf{x} \in \mathbb{R}$, where $\mathbb{R}$ is a designated region of state space. The number of states therein is $W_{\mathbb{R}} = \int_{\mathbb{R}} d\mathbf{x}\,\rho(\mathbf{x})$, and since those states are presumed to be equally probable the entropy of a system restricted to them is simply $S = \log W_{\mathbb{R}}$.

The preceding discussion implies that in order to evaluate S by invoking the EAPP hypothesis, two things must be done: (a) The region $\mathbb{R}$ in which the continuous states are presumed to be equally probable must be identified. In exceptional cases $\mathbb{R}$ may comprise the entire state space, but $\mathbb{R}$ is normally determined by conditions which are intended to exclude unequally probable states. (b) The density of states $\rho(\mathbf{x})$ must be defined or determined so that $W_{\mathbb{R}}$ can be evaluated. It cannot be emphasized too strongly that *the EAPP hypothesis for continuous states remains incomplete, ambiguous, and indeterminate until the functional form of* $\rho(\mathbf{x})$ *is known*. Tasks (a) and (b) are conceptually distinct, so they will be addressed in separate subsections.

3.2.1 Equally Probable Continuous States

The logical rationale for delineating equally probable continuous states is essentially isomorphic to that whereby equally probable discrete states were defined in Sect. 3.1. The correspondence between the two cases consists in the fact that the state variables $\mathbf{x}$ are the continuous analog of the discrete state index k, and as such should be regarded as the labels of the continuous states. However, the analogy is compromised by the fact that the discrete probability p_k has a definite value for the state labeled by the index k, whereas the value of the continuous probability density $p(\mathbf{x})$ for the state labeled by $\mathbf{x}$

depends upon the choice of the state variables $\mathbf{x}$. This dependence is merely an artifact of the fact that $p(\mathbf{x})$ is not itself a probability, and can be avoided simply by dealing with the covariant probability distribution $q(\mathbf{x}) = p(\mathbf{x})/\rho(\mathbf{x})$ rather than $p(\mathbf{x})$ itself. As discussed in Sect. 2.6, $q(\mathbf{x})$ has the same numerical value for any choice of the state variables $\mathbf{x}$, which reflects its covariant significance as the probability of each individual state in the neighborhood of $\mathbf{x}$.

Like the discrete probabilities p_k, which are intrinsic properties of the discrete states and therefore cannot depend explicitly on their arbitrary labels k, the covariant probabilities $q(\mathbf{x})$ are intrinsic properties of the continuous states, and as such they cannot depend explicitly on their labels $\mathbf{x}$. Thus $q(\mathbf{x})$ can only depend on $\mathbf{x}$ implicitly via a functional dependence on some other intrinsic property or properties of the states. If no such properties exist then $q(\mathbf{x})$ must be a constant independent of $\mathbf{x}$. The normalization condition of Eq. (2.18) then implies that $q = 1/W$, where $W = \int d\mathbf{x}\, \rho(\mathbf{x})$, which implies that the EAPP hypothesis is rigorously valid and the entropy is given by $S = \log W$. Unfortunately, systems of practical interest are rarely of this type; their states typically possess multiple intrinsic properties, which makes it necessary to decide which of them q is presumed to depend upon. Just as in the case of discrete states, the energy $E(\mathbf{x})$ is often the only intrinsic property of the states upon which the probabilities seem likely to depend. When such is the case, q is presumed to be of the form $q(\mathbf{x}) = \hat{q}(E(\mathbf{x}))$. This in turn implies that if the system is restricted to the subset $\mathbb{E}$ of $\mathbf{x}$-space for which $E(\mathbf{x}) \cong E$, where E is a particular specified value of the energy, then $q(\mathbf{x}) \cong \hat{q}(E)$ and the EAPP hypothesis is approximately valid subject to that restriction. The entropy of the system is then simply given by $S(E) \cong \log W(E)$, where $W(E)$ is the number of states for which $E(\mathbf{x}) \cong E$; i.e., $W(E) = \int_{\mathbb{E}} d\mathbf{x}\, \rho(\mathbf{x})$.

3.2.2 The Density of States

As mentioned in Sect. 2.6, there is no general theory or procedure whereby the density of states $\rho(\mathbf{x})$ can be systematically determined for any particular system of interest and/or choice of the state variables $\mathbf{x}$. Fortunately, in many situations the functional form of $\rho(\mathbf{x})$ can be inferred from group-invariance arguments [19–21]. Such arguments are somewhat *ad hoc* and are specific to the context, so they must be devised on a case-by-case basis. However, they share the common feature that they exploit whatever symmetry properties the particular system, or class of systems, are perceived to possess. As a trivial example, let $\rho(\theta)$ represent the density of states on the unit circle as a function of the usual polar angle θ, and consider the transformed density of states $\tilde{\rho}(\theta)$ obtained by rotating the function $\rho(\theta)$ counterclockwise by an angle θ_0. Clearly $\tilde{\rho}(\theta) = \rho(\theta - \theta_0)$. If the system is rotationally symmetric then the density of states must be invariant to such rotations, so that $\tilde{\rho}(\theta) = \rho(\theta - \theta_0) = \rho(\theta)$ for all θ_0, which immediately implies that $\rho(\theta)$ is a constant ρ_0 independent of θ. The angle θ is therefore a natural representation of the state of the system. Of course, ρ will no longer remain constant under a nonlinear transformation from θ to an alternative state variable ξ. Such a transformation is performed by setting $\rho(\xi)\, d\xi = \rho(\theta)\, d\theta$, which implies $\rho(\xi) = (d\theta/d\xi)\, \rho_0$.

The continuous states of primary interest in this book are the dynamical states of classical many-particle systems. Such systems have traditionally been described in terms of conservative Hamiltonian dynamics, in which the states are represented by the canonical coordinates and momenta $(\mathbf{q}, \mathbf{p})$ of the particles, and the energy $E(\mathbf{q}, \mathbf{p})$ is the sum of their kinetic and potential energies. In classical statistical mechanics, the density of states $\rho(\mathbf{q}, \mathbf{p})$ is invariably taken to be a constant independent of $(\mathbf{q}, \mathbf{p})$, so that the variables $(\mathbf{q}, \mathbf{p})$ constitute a natural representation. However, the rationale for this choice of $\rho(\mathbf{p}, \mathbf{q})$ often seems somewhat nebulous, perhaps because the canonical formalism obscures the fact that such a choice is even necessary. The concepts involved are more easily understood within the context of the more general dynamical formalism described below. This development will lead to a general condition which the density of states $\rho(\mathbf{x})$ should ideally satisfy for an arbitrary choice of the state variables $\mathbf{x}$, and which thereby determines when those variables constitute a natural representation.

The salient characteristic of a dynamical system is that it does not occupy a fixed state $\mathbf{x}$ for all time but rather spontaneously traces out a time-dependent trajectory in state space. The state occupied by the system at time t is denoted by $\mathbf{x}(t)$. We shall restrict attention to systems for which $\mathbf{x}(t)$ is determined by dynamical equations of motion of the general form

$$\frac{d\mathbf{x}}{dt} = \mathbf{U}(\mathbf{x}) \tag{3.3}$$

where $\mathbf{U}(\mathbf{x})$ is a smooth function of $\mathbf{x}$ which can be interpreted as a stationary velocity field in state space. Equation (3.3) is a system of coupled nonlinear ordinary differential equations which uniquely determine $\mathbf{x}(t)$ in terms of the initial state $\mathbf{x}_0$ at time $t = 0$. Canonical Hamiltonian dynamics is a special case of Eq. (3.3) in which $\mathbf{x} = (\mathbf{q}, \mathbf{p})$ and the equations of motion take the form

$$\frac{d\mathbf{q}}{dt} = \frac{\partial E}{\partial \mathbf{p}} \tag{3.4}$$

$$\frac{d\mathbf{p}}{dt} = -\frac{\partial E}{\partial \mathbf{q}} \tag{3.5}$$

so that $\mathbf{U} = (\partial E/\partial \mathbf{p}, -\partial E/\partial \mathbf{q})$. It has long been traditional to refer to the canonical state space $(\mathbf{q}, \mathbf{p})$ as *phase space*, a terminology which is conventionally extended to apply to an arbitrary dynamical state space $\mathbf{x}$ as well.

A constant of the motion is a state function $A(\mathbf{x})$ whose value does not change with time; i.e.,

$$\frac{dA}{dt} = \frac{\partial A}{\partial \mathbf{x}} \cdot \frac{d\mathbf{x}}{dt} = \mathbf{U} \cdot \nabla A = 0 \tag{3.6}$$

where $\nabla \equiv \partial/\partial\mathbf{x}$ is the gradient operator in $\mathbf{x}$-space. Clearly any function of one or more constants of the motion is also a constant of the motion. Constants of the motion constitute the most important class of intrinsic state properties or variables. It is trivial to verify that $E(\mathbf{q}, \mathbf{p})$ is a constant of the motion in Hamiltonian systems, for which $\nabla E = (\partial E/\partial\mathbf{q}, \partial E/\partial\mathbf{p})$.

Since the state $\mathbf{x}(t)$ of a particular system changes with time, the probability density which describes the uncertainty as to which state a system occupies must likewise change with time in such a way that it tracks and follows the dynamical time evolution of the states. The latter time evolution is determined by the function $\mathbf{U}(\mathbf{x})$, which therefore also determines the time evolution of the probability density $p(\mathbf{x}, t)$ in terms of the initial probability density $p(\mathbf{x}, 0)$. The resulting evolution equation for $p(\mathbf{x}, t)$ is the so-called generalized Liouville equation:

$$\frac{\partial p}{\partial t} + \nabla \cdot (p\mathbf{U}) = 0 \tag{3.7}$$

The proof of Eq. (3.7) is not difficult [25], but will be omitted for brevity. Readers familiar with fluid dynamics will recognize that Eq. (3.7) is simply the continuity equation in state space. The familiar Liouville equation for canonical Hamiltonian systems [26] is a special case of Eq. (3.7).

Depending on the dimensionality of the state space and the form of $\mathbf{U}(\mathbf{x})$, the solutions $\mathbf{x}(t)$ of Eq. (3.3) can exhibit a wide variety of different types of solution behavior. The dynamical systems of present interest are those for which $\mathbf{x}(t)$ executes undamped bounded motion which is confined to a finite region of state space. The state

$\mathbf{x}(t)$ of a particular system then never comes to rest, but this does not imply that the probability density $p(\mathbf{x}, t)$ is likewise unsteady. On the contrary, $p(\mathbf{x}, t)$ must be independent of time for systems in statistical equilibrium, which are precisely the systems of interest in this book. Equation (3.7) then implies that equilibrium probability densities $p(\mathbf{x})$ for such systems must satisfy the stationary Liouville equation

$$\nabla \cdot (p\mathbf{U}) = 0 \tag{3.8}$$

Solutions of Eq. (3.8) are not in general unique, since it is easy to verify that if $p(\mathbf{x})$ is such a solution then so is $A(\mathbf{x})\,p(\mathbf{x})$, where $A(\mathbf{x})$ is any constant of the motion, and conversely that the ratio of any two such solutions is a constant of the motion. The present discussion is restricted to dynamical systems for which Eq. (3.8) possesses smooth solutions, which is not always the case [22].

As discussed in Sect. 2.6, the probability density of equally probable continuous states is $p_0(\mathbf{x}) = (1/W)\,\rho(\mathbf{x})$, which describes a situation in which the uncertainty as to which state $\mathbf{x}$ the system occupies is a maximum. Since the uncertainty of such a situation cannot increase, it must persist in time in order not to spontaneously decrease. This requires $p_0(\mathbf{x})$ to satisfy Eq. (3.8), so that $p_0(\mathbf{x})$ describes a system in statistical equilibrium. It then follows that $\rho(\mathbf{x})$ must satisfy the same equation; i.e.,

$$\nabla \cdot (\rho\mathbf{U}) = 0 \tag{3.9}$$

These considerations suggest that in dynamical systems of the form of Eq. (3.3) for which smooth equilibrium probability densities are observed or presumed to exist, the density of states $\rho(\mathbf{x})$ should be required to be a smooth positive solution of Eq. (3.9). Equation (3.9) can be interpreted as a symmetry condition which requires the density of states $\rho(\mathbf{x})$ to be invariant to the group of time translations [22], and thereby serves as another illustration of the use of symmetry to determine $\rho(\mathbf{x})$.

Equations (3.8) and (3.9) combine to imply that

$$\mathbf{U} \cdot \nabla \left(\frac{p}{\rho} \right) = 0 \tag{3.10}$$

which shows that any equilibrium covariant probability distribution $q = p/\rho$ is a constant of the motion, and must therefore be a function of the intrinsic constants of the motion of the dynamical system of Eq. (3.3). This is a special case of our previous general observation that the dependence of $q(\mathbf{x})$ on $\mathbf{x}$ cannot be explicit but must rather be implicit via a functional dependence on the intrinsic properties of the states. In the present dynamical context, Eq. (3.10) shows that those properties must be constants of the motion. It follows that any equilibrium probability density of the form $p(\mathbf{x}) = A(\mathbf{x})\,\rho(\mathbf{x})$ is a solution of Eq. (3.8), where $A(\mathbf{x})$ is a constant of the motion. This reemphasizes that the equilibrium probability density is not unique.

Although Eq. (3.9) has the same form as Eq. (3.8), $\rho(\mathbf{x})$ is subject to a subtle but essential restriction which does not apply to $p(\mathbf{x})$, and which combines with Eq. (3.9) to imply that $\rho(\mathbf{x})$ is unique to within a constant multiplicative factor. That restriction arises from the fact that $\rho(\mathbf{x})$ is a primitive quantity which simply represents the density of states in $\mathbf{x}$-space *and nothing more*; i.e., it has no additional interpretation or significance. Thus, by its very nature $\rho(\mathbf{x})$ is inherently unrelated to whatever constants of the motion $\mathbf{A}(\mathbf{x})$ or other intrinsic properties the states themselves may or may not possess. It would therefore be conceptually incongruous and inconsistent to allow $\rho(\mathbf{x})$ to possess a functional dependence upon such properties. Thus we impose the restriction that any proper permissible density of states $\rho(\mathbf{x})$ must be functionally independent of constants of the motion.

The uniqueness proof then proceeds as follows. Let $\rho_1(\mathbf{x})$ and $\rho_2(\mathbf{x})$ denote two solutions of Eq. (3.9), so that $\nabla\cdot(\rho_1\mathbf{U}) = \nabla\cdot(\rho_2\mathbf{U}) = 0$. Then $\nabla\cdot[(\rho_2/\rho_1)\rho_1\mathbf{U}] = \rho_1\mathbf{U}\cdot\nabla(\rho_2/\rho_1) = 0$, which implies that ρ_2/ρ_1 is a constant of the motion, call it $A(\mathbf{x})$. Let us suppose that

$\rho_1(\mathbf{x})$ is a proper density of states which is functionally independent of any constants of the motion. This then implies that $\rho_2(\mathbf{x}) = A(\mathbf{x})\,\rho_1(\mathbf{x})$ is functionally dependent on $A(\mathbf{x})$, and is therefore not a proper density of states (except of course in the trivial case when $A(\mathbf{x})$ is a pure constant independent of $\mathbf{x}$). But $\rho_2(\mathbf{x})$ represents any solution of Eq. (3.9) other than $\rho_1(\mathbf{x})$, so no such solution can be a proper density of states. It follows that $\rho_1(\mathbf{x})$ is in fact the only proper density of states which is functionally independent of the constants of the motion, and is therefore unique to within an indeterminate multiplicative constant. According to Eq. (2.19), such a constant corresponds to an indeterminate addititve constant in the BGS entropy S, and thereby in the BP entropy as well when the states are equally probable. This ambiguity does not occur for systems with discrete states, but in any case it is immaterial in situations where only entropy differences are of interest, as is usually the case. In a specific system or context, however, it may be useful to resolve this ambiguity by requiring S to assume a desired value as a special or limiting case. In particular, in the classical statistical mechanics of thermodynamic systems it is convenient to choose the indeterminate constant factor in $\rho(\mathbf{x})$ so that S is consistent with the classical limit of quantum mechanics. The manner in which this is done will be discussed in Chapter 8.

Alas, the preceding uniqueness proof does not supply the functional form of $\rho(\mathbf{x})$. The analytical determination of $\rho(\mathbf{x})$ is not tractable in general, but it becomes trivial in the important special case when the velocity field $\mathbf{U}(\mathbf{x})$ is incompressible; i.e., when

$$\nabla \cdot \mathbf{U} = 0 \tag{3.11}$$

Equation (3.9) then reduces to $\mathbf{U} \cdot \nabla\rho = 0$, which shows that $\rho(\mathbf{x})$ itself is a constant of the motion. As discussed above, however, $\rho(\mathbf{x})$ cannot depend on nontrivial constants of the motion, so it must simply be a constant ρ_x independent of $\mathbf{x}$; i.e.,

$$\rho(\mathbf{x}) = \rho_x \tag{3.12}$$

where again the subscript x is not a variable but merely a label which serves to identify the set of variables $\mathbf{x}$. The actual value of the constant ρ_x is indeterminate and is usually immaterial as discussed above. Conversely, Eqs. (3.9) and (3.12) combine to imply Eq. (3.11). Thus the state variables $\mathbf{x}$ constitute a natural representation when Eq. (3.11) is satisfied, and vice versa. Canonical Hamiltonian dynamics constitutes a special case of this situation in which the equations of motion are Eqs. (3.4) and (3.5), $\mathbf{x} = (\mathbf{q}, \mathbf{p})$, $\nabla = (\partial/\partial\mathbf{q}, \partial/\partial\mathbf{p})$, and $\mathbf{U} = (\partial E/\partial\mathbf{p}, -\partial E/\partial\mathbf{q})$, so that $\nabla \cdot \mathbf{U} = (\partial^2 E/\partial\mathbf{q}\partial\mathbf{p}) - (\partial^2 E/\partial\mathbf{p}\partial\mathbf{q}) = 0$. Thus we see that *the fundamental basis and rationale for setting the density of states* $\rho(\mathbf{q}, \mathbf{p})$ *equal to a constant in Hamiltonian systems derives from the more general Eq. (3.9)*, which thereby justifies and confirms the usual procedure for invoking the EAPP hypothesis in canonical phase space [22, 26].

Equations (3.11) and (3.12) are not covariant with respect to an arbitrary nonlinear transformation $\mathbf{y} = \mathbf{f}(\mathbf{x})$ of the state variables, which converts Eq. (3.3) into the equivalent dynamical system $d\mathbf{y}/dt = \mathbf{V}(\mathbf{y})$, where $\mathbf{V}(\mathbf{y}) \equiv \mathbf{U}(\mathbf{x})\cdot(\partial\mathbf{f}/\partial\mathbf{x})$. This observation suggests that if Eq. (3.11) is not already satisfied for a particular set of variables $\mathbf{x}$, perhaps a transformation could be found whereby it is satisfied for the new variables $\mathbf{y}$; i.e., $(\partial/\partial\mathbf{y})\cdot\mathbf{V}(\mathbf{y}) = 0$. If so, it would then follow that $\rho(\mathbf{y})$ is a constant ρ_y independent of $\mathbf{y}$, so that the new variables $\mathbf{y}$ constitute a natural representation. However, $\rho(\mathbf{x}) = J\,\rho(\mathbf{y})$, where $J = \|\partial\mathbf{f}/\partial\mathbf{x}\|$ is the Jacobian of the transformation. Thus $\rho(\mathbf{x}) = \rho_y\, J(\mathbf{x})$, so solving for J is no more tractable than solving Eq. (3.9) for $\rho(\mathbf{x})$. This approach therefore seems unlikely to be useful, except perhaps in special cases where the required transformation can be inferred by inspection.

It is of interest to examine the nonequilibrium time dependence of the BGS entropy S implied by Eq. (3.7). Combining Eqs. (2.19) and (3.7), we obtain

$$\frac{dS}{dt} = \int d\mathbf{x}\, p(\mathbf{x}, t)\, D(\mathbf{x}) \tag{3.13}$$

where $D(\mathbf{x}) = (1/\rho)\,\nabla \cdot (\rho \mathbf{U})$ is the covariant divergence of the velocity field $\mathbf{U}(\mathbf{x})$ with respect to the volume element $\rho(\mathbf{x})\,d\mathbf{x}$ in state space, and two integrations by parts have been performed. However, Eq. (3.9) shows that $D = 0$, so that $dS/dt = 0$ as well. The entropy $S(t)$ at time t is therefore independent of t and remains constant at its initial value $S(0)$, in spite of the fact that the system is not in statistical equilibrium; i.e., $\partial p/\partial t \neq 0$. Note that this result is valid for an arbitrary initial probability density $p(\mathbf{x}, 0)$. The uncertainty as to which state the system occupies is therefore invariant to time translations even when it is not initially a maximum and the probability density $p(\mathbf{x}, t)$ is continuing to evolve in time. This invariance may superficially seem inconsistent with the second law of thermodynamics, according to which the entropy would be expected to increase with time. This apparent paradox will be resolved in Sect. 5.4.

Chapter 4

CONSTRAINTS

In most situations of physical interest, the number of accessible states W and their probabilities p_k are not fixed constants but rather depend on the values of a relatively small number of externally controlled parameters. Such parameters are typically associated with restrictions imposed on the system in order to (a) ensure that the number of accessible states is finite, as discussed in Sect. 2.7, and/or (b) confine the system to a subset of states for which the EAPP hypothesis is presumed to be valid, as discussed in Chapter 3. In this chapter we shall presume that conditions (a) and (b) are both satisfied, so that the number of accessible states W is finite and their probabilities p_k are equal with the common value $p_k = 1/W$. The entropy of the system is then the BP entropy $S = \log W$. The functional dependence of W and S on whatever parameters are associated with conditions (a) and (b) is understood and must be remembered, but in this chapter it will be suppressed from the notation for simplicity. The purpose of this chapter is to develop a suitable mathematical framework to explore and describe the implications of imposing additional constraints on the system. The resulting constrained values of W and S then become functions of the parameters associated with those constraints. The leitmotiv of the discussion is that those functions conversely determine the probability distributions or densities of those constraint parameters and their fluctuations in the unconstrained system. The resulting expressions for the probability densities are generalizations of those first used by Einstein to describe thermodynamic fluctuations.

4.1 Constraints on Subsets of States

In many situations a set of equally probable discrete states $k = 1, 2, \cdots, W$ is naturally divided into mutually exclusive subsets or groups by certain parameters which describe the configuration of the system. The resulting subsets of states can then be regarded as the basic units of which the entire state space is composed. To describe such situations, we suppose that there are M such subsets which are labeled by an index $\nu = 1, 2, \cdots, M$. The number of states in subset ν is denoted by W_ν. The subsets are presumed to be exhaustive, so that

$$W = \sum_{\nu=1}^{M} W_\nu \tag{4.1}$$

Thus W can be determined by first evaluating the quantities W_ν and then summing over ν. Now if some of the states are removed from any set of equally probable states, the remaining states clearly remain equally probable. The states within subset ν are therefore equally probable, because they are the states that remain behind if one removes all the states in all of the other subsets. The entropy of subset ν is therefore its BP entropy $S_\nu = \log W_\nu$. It is essential to observe that S_ν can also be interpreted as the constrained entropy of the entire system subject to the constraint that the state of the system lies in subset ν. Since $W_\nu \leq W$, it follows that $S_\nu \leq S$. Thus we see that constraining a system of equally probable states to a subset of those states cannot increase its entropy, and will ordinarily reduce it. Conversely, the entropy cannot decrease, and will ordinarily increase, when such a constraint is removed. This provides our second glimpse of the principle of maximum entropy, which in this context is an essentially trivial consequence of the fact that the BP entropy $S = \log W$ decreases when the number of accessible states W is reduced by imposing a constraint, and conversely increases when W is increased by removing a constraint.

It is important to note, however, that there is a fundamental asymmetry between imposing and removing constraints. Consider the effect of first imposing the constraint that the state of the system lies within a particular subset ν, and then subsequently removing or relaxing that constraint. Imposing the constraint immediately reduces the number of accessible states from W to $W_\nu < W$, thereby reducing the entropy from $S = \log W$ to $S_\nu = \log W_\nu < \log W = S$. Conversely, when the constraint is removed the system is once again free to occupy states in the other subsets. However, it will actually do so only if some mechanism exists for producing spontaneous transitions (either random or deterministic) between states. Moreover, in order for that mechanism to restore the original values of W and S it must clearly provide access to all or almost all of the states k, which loosely speaking is the essence of ergodicity. In the absence of such a mechanism, the system will be unaffected by removing the constraint and will simply remain in the state k within subset ν to which it was consigned when the constraint was imposed. The value of the entropy then remains S_ν and does not increase when the constraint is removed, consistent with the fact that there has been no increase in the uncertainty as to which state the system occupies.

A simple example may make the asymmetry of the situation easier to visualize. Consider a single pawn on a chessboard which is concealed from view. The pawn is the system and the squares are the states, so $W = 64$ and $S = \log 64$. Now suppose the constraint is imposed that the pawn is forbidden to occupy a light square. Thus, if it were initially on a light square the constraint would summarily evict it and force it to move to one of the 32 dark squares, thereby reducing its entropy to $S = \log 32$. But once this has been done, simply rescinding or repealing the constraint does not force the pawn to move again. It will therefore remain on its current dark square until and unless something (e.g., mechanical vibrations of the chessboard, or children playing with it) causes it to move again, thereby restoring access to the light squares and causing W and S to increase back to their original values. In thermodynamic systems, the mechanism that produces spontaneous transitions between microstates is thermal agitation of the constituent particles.

Consider again the effect of imposing the constraint that the state of the system lies within a particular subset ν. The increase in entropy which results when that constraint is removed is simply $\Delta S_\nu \equiv \log W - \log W_\nu = -\log(W_\nu/W)$. The weighted average of ΔS_ν over all subsets ν, weighted by the number of states in each subset, is then given by $\Delta S \equiv -\sum_\nu (W_\nu/W)\log(W_\nu/W) = -\sum_\nu p_\nu \log p_\nu$. Thus ΔS is identical to the BGS entropy associated with the uncertainty as to which subset ν the state of the system lies in. This is eminently sensible, since it is precisely that uncertainty which is removed, on the average, when the state of the system is constrained to lie within a particular subset ν, and which conversely is introduced when such a constraint is removed.

It is also noteworthy that the quantity ΔS is isomorphic to the familiar *entropy of mixing*, which is an important concept that naturally arises in connection with the physical mixing of two or more different substances. It therefore lies outside the scope of this book as defined in the Introduction, but its structural resemblance to ΔS reflects an intimate relation between the two concepts. That relation lends a great deal of insight into the significance of the entropy of mixing, so we would be remiss not to discuss it. The essence of the matter is that a mixture of different substances can be regarded as a composite system of which each individual substance constitutes a subsystem. Those subsystems can then be subjected to constraints on the states they are allowed to occupy. In particular, if the substances are initially separated from one another by impermeable membranes, then each substance is thereby subjected to the constraint that it is denied access to the spatial states occupied by the other substances. When those constraints are removed the substances become free to occupy those previously inaccessible spatial states, and in doing so they mix together. At the same time, the entropy of the system increases due to the increase in the number of states that have thereby become accessible to each of the substances. The resulting increase in entropy has long been referred to as the "entropy of mixing." Ironically, it has little or nothing to do with mixing per se [27, 28]; it should rather be properly interpreted as a mere special case of the increase in the entropy that generally occurs when constraints are removed, as discussed above.

Equation (4.1) can be rewritten in terms of the entropies S and S_ν as

$$\exp S = \sum_{\nu=1}^{M} \exp S_\nu \tag{4.2}$$

which shows that even though S is additive over statistically independent subsystems by construction, it is *not* likewise additive over constrained subsets of states. Since the states k are equally probable, the probability that the state of the unconstrained system lies in subset ν is simply

$$p_\nu = \frac{W_\nu}{W} = \frac{\exp S_\nu}{\exp S} = \exp(S_\nu - S) \tag{4.3}$$

Thus the probability that the state of an *unconstrained* system lies within a particular subset is proportional to the exponential of its entropy when its state is *constrained* to lie within that subset. This superficially trivial relation between probabilities and exponentials of constrained entropies is actually a general feature of fundamental importance, as will be seen in what follows.

The value of ν for which p_ν attains its maximum value p^* is denoted by ν^*, and is presumed to be unique for simplicity. Thus the index ν^* defines the most probable subset of states. According to Eq. (4.3), W_ν and S_ν also attain their maximum values $W^* = p^* W$ and $S^* = \log W^*$ when $\nu = \nu^*$, so that $p^* = \exp(S^* - S)$ and $S^* = S + \log p^*$. Clearly $W^* \leq W$ and $S^* \leq S$. It is often easier to determine W^* or S^* than it is to perform the summation necessary to determine W or S via Eq. (4.1) or (4.2), so it is of interest to examine the conditions under which W and S can be accurately approximated by W^* and S^*, respectively. The approximation $S \cong S^*$ is of direct practical importance and utility, because when it is valid S can be evaluated simply by maximizing S_ν with respect to ν; i.e., $S \cong \max_\nu S_\nu$. This is yet another manifestation of the principle of maximum entropy (PME), albeit a conditional one. It must be emphasized that the PME should not be misconstrued as asserting or implying that the approximation $S \cong S^*$ is universally valid, which is most emphatically not the case.

It is clear that the approximation $W \cong W^*$ is accurate only if $W - W^* \ll W$, which implies that the most probable subset ν^* contains almost all of the W states of the entire system. As discussed at the end of Sect. 2.3, this condition is equivalent to

$$\log(W/W^*) \ll 1 \tag{4.4}$$

and when it is satisfied it clearly implies that $S \cong S^*$ as well. However, Eq. (4.4) is stronger than necessary to ensure the accuracy of the approximation $S \cong S^*$, which merely requires that $\log W - \log W^* \ll \log W$, or

$$\log(W/W^*) \ll \log W \tag{4.5}$$

Equation (4.5) is much less restrictive than Eq. (4.4) when W is large, so it is unnecessary for the most probable subset ν^* to contain almost all the states in order for S^* to accurately approximate S. Equation (4.5) roughly determines the minimum fraction W^*/W of the states which the most probable subset must contain in order for the approximation $S \cong S^*$ to be valid. When W is very large, this fraction can be very small (i.e., W/W^* and even $\log(W/W^*)$ can be very large) without invalidating the approximation $S \cong S^*$. When such is the case, the entropy $S \cong S^* = \log W^*$ can be accurately evaluated by counting only a very small fraction or percentage of the total number of states, which is a paradoxical and counterintuitive consequence of taking logarithms of very large numbers. Logarithms forgive many sins, whereas exponentials magnify them.

The smallest possible value of W^* is the mean number of states per subset W/M, so $W^* > W/M$ and $\log(W/W^*) < \log M$. The condition

$$\log M \ll \log W \tag{4.6}$$

is therefore always sufficient to imply that Eq. (4.5) is satisfied, and thereby to ensure that the approximation $S \cong S^*$ is accurate. It is instructive to confirm this condition by another method based on the so-called squeeze technique, wherein a quantity Q of interest is determined by first showing that $Q_1 \leq Q \leq Q_2$ and then showing that $Q_1 \cong Q_2$, so that $Q \cong Q_1 \cong Q_2$. The application of this technique in the present context is based on the observation that Eq. (4.1) immediately implies $W^* \leq W \leq M W^*$, so that $S^* \leq S \leq S^* + \log M$, which collapses to $S \cong S^*$ when Eq. (4.6) is satisfied. Note that Eq. (4.6) does *not* require p^* to be much larger than the probabilities of the other subsets $\nu \neq \nu^*$. Indeed, when Eq. (4.6) is satisfied the approximation $S \cong S^*$ remains valid even if all the values of W_ν are very nearly equal to W/M, so that the probabilities p_ν are all very nearly equal to $1/M$. Equation (4.6) therefore ensures that $S \cong S^*$ by brute force, as it were; it simply requires the number of subsets M to be small enough that even the smallest possible value of W^*, namely W/M, is large enough that $\log W^* \cong \log W$. Equation (4.6) is therefore much more restrictive than necessary in situations where $W^* \gg W/M$, which indeed are quite common. In such situations, the W states of the system are necessarily distributed amongst its M subsets in a highly nonuniform manner, and it may happen that they are so densely concentrated in a smaller number $M_\circledast$ of subsets (which of course include the most probable subset ν^*, as the subscript $\circledast$ is meant to suggest) that the number of states in the remaining $M - M_\circledast$ subsets can be neglected. When such is the case, those $M - M_\circledast$ relatively empty subsets can be omitted from consideration, which effectively reduces the number of subsets from M to $M_{\circledast}$. Under these circumstances M can be replaced by $M_{\circledast}$ in Eq. (4.6), which then becomes much less restrictive when $M_{\circledast} \ll M$. That is to say, if almost all of the W states of the system are concentrated in $M_{\circledast} < M$ subsets, and $M_{\circledast}$ is sufficiently small that $\log M_{\circledast} \ll \log W$, then the approximation $S \cong S^*$ is valid.

Equation (4.3) shows that the probabilities p_ν can be computed from either the quantities W_ν or the subset entropies S_ν. In the latter case, however, it must be kept in mind that accuracy of S_ν merely requires logarithmic accuracy of W_ν, and conversely that relatively small errors in S_ν can be magnified into relatively large errors in W_ν by the exponential in Eq. (4.3). Thus accurate values of the quantities S_ν are in general insufficient to ensure accurate values of the W_ν. The accuracy of the probabilities p_ν obtained from Eq. (4.3) is thereby called into question. Fortunately, this concern is ameliorated by the fact that errors in W_ν produce corresponding errors in W, which tend to partially cancel out when W_ν is divided by W to obtain p_ν. This cancellation of errors is in fact nearly complete when all the quantities W_ν that contribute significantly to W are in error by nearly the same factor, in which case the corresponding probabilities $p_\nu = W_\nu/W$ are accurate in spite of the errors in W_ν, however large they may be. In practice, this happy situation normally occurs only when the probabilities p_ν are sharply peaked in the sense that $p^* \gg 1/M$. Such is often the case for large W, but of course this cannot simply be assumed without justification.

4.2 Constraints on State Variables

Another common and useful class of constraints consists of those in which the discrete states $k = 1, 2, \cdots, W$ themselves are not constrained directly as they were in the preceding section, but rather indirectly by restricting the allowed values $\mathbf{A}_k$ of a specified set of state variables or observables $\mathbf{A} = (A_1, A_2, \cdots)$. Two types of such constraints must be distinguished, which will be referred to as *sharp* (or strong) constraints and *mean* (or weak) constraints, both of which have the effect of altering the probabilities of the states. Since the states $k = 1, 2, \cdots, W$ are equally probable prior to imposing either type of constraints, their initial probabilities are simply $p_k = 1/W$, independently of k, and the entropy of the unconstrained system is again the BP entropy $S = \log W$.

4.2.1 Sharp Constraints

We define a sharp constraint as one whereby the system is allowed to occupy only states k for which $\mathbf{A}_k$ has a definite specified value $\mathbf{a} = (a_1, a_2, \cdots)$. Those states then remain equally probable, while all other states are rendered inaccessible, so the entropy of the constrained system is simply the BP entropy; i.e., the logarithm of the number of accessible states. As briefly discussed in Sect. 3.1, however, the number of states k that precisely satisfy the condition $\mathbf{A}_k = \mathbf{a}$ is almost always zero, because the values $\mathbf{A}_k$, like the states k themselves, are discrete, whereas $\mathbf{a}$ is a continuous variable. The number of states k for which $\mathbf{A}_k = \mathbf{a}$ is therefore zero for almost all values of $\mathbf{a}$, in much the same way that almost all real numbers are irrational. The logarithm of zero diverges, so the BP entropy is not well defined as a function of $\mathbf{a}$. It follows that *infinitely precise constraints cannot be imposed on state variables in systems with discrete states.* Indeed, precise constraints cannot be imposed on systems with continuous states either. As discussed in Sect. 2.6, the number of states within any region $\mathbb{R}$ of a continuous state space $\mathbf{x}$ is simply $W_{\mathbb{R}} = \int_{\mathbb{R}} d\mathbf{x}\, \rho(\mathbf{x})$, where $\rho(\mathbf{x})$ is the density of states. But the region $\mathbb{R}$ to which $\mathbf{x}$ is restricted by the condition $\mathbf{A}(\mathbf{x}) = \mathbf{a}$ is a hypersurface of lower dimensionality than $\mathbf{x}$. Such a region is a set of measure zero in $\mathbf{x}$-space, so that $W_{\mathbb{R}} = 0$.

To circumvent this dilemma and obtain sensible results, it is again necessary to incorporate a small tolerance or uncertainty into the definition of a sharp constraint. To this end, we replace the condition $\mathbf{A}_k = \mathbf{a}$ by the more tolerant constraint $\mathbf{A}_k \cong \mathbf{a}$, which will henceforth be interpreted as merely requiring $\mathbf{A}_k$ to lie within a small but finite neighborhood $\mathbb{N}(\mathbf{a})$ of $\mathbf{a}$-space which contains the specified value $\mathbf{a}$. This requirement can be enforced in various ways, including the obvious pedestrian approach of conceptually subdividing $\mathbf{a}$-space into a large but finite number of small fixed cells, so that each cell defines the neighborhood of the points that lie within it.† However, it is both simpler and more elegant to judiciously refrain from defining $\mathbb{N}(\mathbf{a})$ any more precisely until the need arises. Thus we simply replace the untenable precise constraint $\mathbf{A}_k = \mathbf{a}$ by

$$\mathbf{A}_k \in \mathbb{N}(\mathbf{a}) \tag{4.7}$$

In the present context, the states k constitute the microstates of the system, while the variables $\mathbf{a}$ define the macrostates.

The number of microstates k consistent with Eq. (4.7) will be denoted by $W(\mathbf{a})$. As previously discussed, those states remain equally probable, while the remaining states become inaccessible. The entropy of the constrained system as a function of $\mathbf{a}$ is therefore simply the BP entropy

$$S(\mathbf{a}) \ = \ \log W(\mathbf{a}) \tag{4.8}$$

and it is obvious that $W(\mathbf{a}) < W$ and $S(\mathbf{a}) < S$. In principle $W(\mathbf{a})$ is an integer, which implies that $W(\mathbf{a})$ and consequently $S(\mathbf{a})$ are discontinuous functions of $\mathbf{a}$. However, the present formulation is useful only when $W(\mathbf{a})$ and $S(\mathbf{a})$ can be accurately approximated by continuous functions, which requires the discontinuities in $W(\mathbf{a})$ to be much smaller than $W(\mathbf{a})$ itself. Thus we restrict attention to systems with $W \gg 1$, and we require the neighborhoods $\mathbb{N}(\mathbf{a})$ to be large enough that

$$W(\mathbf{a}) \ \gg \ 1 \tag{4.9}$$

---

†Note that a-space is finite because we have previously contrived to ensure that the number of accessible states is finite, which in turn implies that the maximum and minimum values of the observables $\mathbf{A}$ are likewise finite.

so that $W(\mathbf{a})$ and $S(\mathbf{a})$ can be regarded and treated as continuous functions of $\mathbf{a}$ for practical purposes. At the same time, we must require the neighborhoods $\mathbb{N}(\mathbf{a})$ to remain small enough that Eq. (4.7) retains its desired significance as a sharp constraint. The latter condition can be expressed as

$$v(\mathbf{a}) \ll V_a \tag{4.10}$$

where

$$v(\mathbf{a}) \equiv \int_{\mathbb{N}(\mathbf{a})} d\mathbf{a}' \tag{4.11}$$

is the volume of the neighborhood $\mathbb{N}(\mathbf{a})$, and $V_a \equiv \int d\mathbf{a}$ is the total volume of $\mathbf{a}$-space (which as noted above is finite). We further require that both $v(\mathbf{a})$ and $W(\mathbf{a})$ vary slowly with $\mathbf{a}$ in the sense that $v(\mathbf{a}') \cong v(\mathbf{a})$ and $W(\mathbf{a}') \cong W(\mathbf{a})$ if $\mathbf{a}' \in \mathbb{N}(\mathbf{a})$ or $\mathbf{a} \in \mathbb{N}(\mathbf{a}')$. If this requirement cannot be satisfied without violating Eq. (4.9) then the present formulation is inapplicable.

Equation (4.9) requires the number of microstates $W(\mathbf{a})$ in the neighborhood $\mathbb{N}(\mathbf{a})$ to be very large, while Eq. (4.10) requires the volume $v(\mathbf{a})$ which contains them to be very small. These two requirements combine to imply a restriction to systems in which the values of $\mathbf{A}_k$ are very closely spaced and densely distributed throughout $\mathbf{a}$-space. The quantities $\mathbf{A}_k$ can then be collectively visualized as a quasi-continuous dense cloud of points in $\mathbf{a}$-space. In such systems $v(\mathbf{a})$ is the only relevant property of $\mathbb{N}(\mathbf{a})$, so it is unnecessary to define $\mathbb{N}(\mathbf{a})$ any more precisely. As will be seen in Chapter 6, these conditions are typically satisfied for macroscopic state variables in thermodynamic systems, for which $S(\mathbf{a})$ typically exhibits a plateau value which is insensitive to $v(\mathbf{a})$ within very wide limits. However, this behavior cannot be assumed *a priori* but requires demonstration.

The probability density $p(\mathbf{a})$ in $\mathbf{a}$-space is defined so that $p(\mathbf{a})\, d\mathbf{a}$ is the probability that the *unconstrained* system occupies a microstate k for which $\mathbf{A}_k$ lies within an infinitesimal volume element $d\mathbf{a}$ containing the macrostate $\mathbf{a}$. Consider an arbitrary region $\mathbb{R}$ of $\mathbf{a}$-space, and let $W_{\mathbb{R}}$ be the number of microstates k for which $\mathbf{A}_k \in \mathbb{R}$. The probability that the system occupies one of those microstates is then $\int_{\mathbb{R}} d\mathbf{a}\, p(\mathbf{a}) = W_{\mathbb{R}}/W$. The true discontinuous probability density is therefore $p(\mathbf{a}) = (1/W) \sum_k \delta(\mathbf{a} - \mathbf{A}_k)$. Since the microstates k are densely distributed in a-space, however, it is more convenient to define and work with a smoothed or mean probability density $\bar{p}(\mathbf{a})$ which is a continuous approximation to $p(\mathbf{a})$ in the sense that

$$\int_{\mathbb{R}} d\mathbf{a}\; \bar{p}(\mathbf{a}) \;\cong\; \int_{\mathbb{R}} d\mathbf{a}\; p(\mathbf{a}) \;=\; \frac{W_{\mathbb{R}}}{W} \tag{4.12}$$

for any region $\mathbb{R}$ which is large enough that $W_{\mathbb{R}} \gg 1$. Note that $p(\mathbf{a})$ and hence $\bar{p}(\mathbf{a})$ depend only on how the points $\mathbf{A}_k$ are distributed in a-space, and are entirely unrelated to and independent of the neighborhoods $\mathbb{N}(\mathbf{a})$. Setting $\mathbb{R} = \mathbb{N}(\mathbf{a})$ in Eq. (4.12), we obtain

$$\int_{\mathbb{N}(\mathbf{a})} d\mathbf{a}'\; \bar{p}(\mathbf{a}') \;=\; \frac{W(\mathbf{a})}{W} \tag{4.13}$$

Under the conditions imposed above, $\bar{p}(\mathbf{a})$ can evidently be defined to be slowly varying in the same sense as $W(\mathbf{a})$, whereupon Eq. (4.13) reduces to

$$\bar{p}(\mathbf{a})\, v(\mathbf{a}) \;=\; \frac{W(\mathbf{a})}{W} \tag{4.14}$$

Since $\bar{p}(\mathbf{a})$ must be normalized so that $\int d\mathbf{a}\, \bar{p}(\mathbf{a}) = 1$, Eq. (4.14) implies that

$$W \;=\; \int d\mathbf{a}\; g(\mathbf{a})\, W(\mathbf{a}) \tag{4.15}$$

where

$$g(\mathbf{a}) \;\equiv\; \frac{1}{v(\mathbf{a})} \tag{4.16}$$

Combining Eqs. (4.8) and (4.14)–(4.16), we obtain

$$\exp S \;=\; \int d\mathbf{a}\; g(\mathbf{a})\, \exp S(\mathbf{a}) \tag{4.17}$$

and

$$\bar{p}(\mathbf{a}) \;=\; g(\mathbf{a}) \exp\{S(\mathbf{a}) - S\} \;=\; \frac{g(\mathbf{a}) \exp S(\mathbf{a})}{\int d\mathbf{a}'\; g(\mathbf{a}') \exp S(\mathbf{a}')} \tag{4.18}$$

Equation (4.18) is analogous to Eq. (4.3) of the previous section, and constitutes an important fundamental relation between the mean probability density $\bar{p}(\mathbf{a})$ in the unconstrained system and the entropy $S(\mathbf{a})$ of the constrained system. The probability density can be used to compute the mean or average value of an arbitrary function $f(\mathbf{a})$, which is given by $\langle f(\mathbf{a})\rangle \equiv \int d\mathbf{a}\,\bar{p}(\mathbf{a})f(\mathbf{a})$, of which fluctuations such as $\langle|\mathbf{a} - \langle\mathbf{a}\rangle|^2\rangle$ are a special case. Relations of this general type were first obtained by Einstein, and are customarily referred to as Einstein fluctuation formulae. They have usually been specifically formulated to describe thermodynamic fluctuations, but the present development shows that they are of much greater generality and are not intrinsically thermodynamical in character. It should be noted, however, that Eq. (4.18) does not yet provide an unambiguous relation between $\bar{p}(\mathbf{a})$ and $S(\mathbf{a})$, because $g(\mathbf{a}) = 1/v(\mathbf{a})$ depends on the volume of the neighborhood $\mathbb{N}(\mathbf{a})$, which has not been precisely specified. This simply reflects the fact that $S(\mathbf{a})$ itself depends upon $v(\mathbf{a})$ via $W(\mathbf{a})$, and hence is ambiguous until $v(\mathbf{a})$ has been defined or determined.

The volume $v(\mathbf{a})$ of the neighborhood $\mathbb{N}(\mathbf{a})$ in $\mathbf{a}$-space is a natural measure of the local tolerance associated with the constraint of Eq. (4.7). Since $v(\mathbf{a})$ is the volume of a single neighborhood, its reciprocal $g(\mathbf{a})$ can be interpreted as the local number density of mutually exclusive (i.e., disjoint) neighborhoods per unit volume in $\mathbf{a}$-space. The total number of such neighborhoods in all of $\mathbf{a}$-space is then $M \equiv \int d\mathbf{a}\,g(\mathbf{a})$, and the mean volume of a single neighborhood is $\bar{v} \equiv V_a/M$. Equation (4.10) then implies $\bar{v} \ll V_a$ and $M \gg 1$. The parameter $\bar{v}/V_a = 1/M \ll 1$ thereby serves as a dimensionless measure of the mean or global tolerance associated with the constraint of Eq. (4.7).

The question naturally arises as to whether the formulation might be simplified by defining $v(\mathbf{a})$, or equivalently $g(\mathbf{a})$, to be a constant independent of $\mathbf{a}$. The neighborhoods $\mathbb{N}(\mathbf{a})$ would then all have the same volume in $\mathbf{a}$-space, thereby making the tolerance associated with Eq. (4.7) independent of $\mathbf{a}$, which superficially seems sensible and desirable. This choice has indeed been adopted in much of the literature, often without comment. Unfortunately, this simple and superficially attractive expedient is untenable as a general *ansatz* because the resulting formulation is not properly covariant with respect to smooth nonlinear transformations of the state variables $\mathbf{A}$. This important point is rarely discussed and often overlooked, and warrants a detailed analysis to which we now turn our attention.

4.2.2 Covariance Considerations

The essence of covariance is that there is nothing unique or special about the variables $\mathbf{A}$ because an alternative equivalent set of variables $\mathbf{B}$ can always be defined by letting $\mathbf{B} = \mathbf{F}(\mathbf{A})$, where $\mathbf{F}(\mathbf{a})$ is an arbitrary invertible nonlinear function of $\mathbf{a}$. The variables $\mathbf{B}$ have the definite values $\mathbf{B}_k = \mathbf{F}(\mathbf{A}_k)$ in state k, so they too are state variables to which the preceding formalism can equally well be applied. The entropy cannot be allowed to depend upon which equivalent set of variables is used to specify or parameterize the same constraints, so we must require that

$$S(\mathbf{b}) = S(\mathbf{a}) \tag{4.19}$$

where $\mathbf{b} = \mathbf{F}(\mathbf{a})$. It is understood that this and similar equations which relate the same quantities as functions of $\mathbf{a}$ and $\mathbf{b}$ are to be interpreted according to the usual loose functional notation common in physics, in which $S(\mathbf{a})$ denotes the actual value of the quantity S for a given value of $\mathbf{a}$ rather than the form of the function of $\mathbf{a}$ which returns the value S. By mathematical standards this notation constitutes an egregious *abus de langage*; it would be more precise to write $S = S_A(\mathbf{a}) = S_B(\mathbf{b})$, thereby distinguishing the two different functional forms by subscripts, but that notation rapidly becomes so cumbersome that suppressing such subscripts is the lesser evil. Since $S(\mathbf{a}) = \log W(\mathbf{a})$, Eq. (4.19) requires $W(\mathbf{b}) = W(\mathbf{a})$, which can be ensured by requiring that $\mathbb{N}(\mathbf{b}) = \mathbb{N}(\mathbf{a})$ so that Eq. (4.7) defines precisely the same group of states k regardless of whether it is written in terms of the variables $\mathbf{a}$ or $\mathbf{b}$. That is to say, the neighborhoods

$\mathbb{N}(\mathbf{b})$ in b-space are simply defined by applying the transformation $\mathbf{b} = \mathbf{F}(\mathbf{a})$ to the neighborhoods $\mathbb{N}(\mathbf{a})$ in a-space. Thus $\mathbf{b}' \in \mathbb{N}(\mathbf{b})$ if and only if $\mathbf{a}' \in \mathbb{N}(\mathbf{a})$, where $\mathbf{b}' = \mathbf{F}(\mathbf{a}')$.

The differential probabilities $\bar{p}(\mathbf{b})\, d\mathbf{b}$ and $\bar{p}(\mathbf{a})\, d\mathbf{a}$ must of course be equal, so

$$\bar{p}(\mathbf{b}) = J^{-1}\, \bar{p}(\mathbf{a}) \tag{4.20}$$

where

$$J(\mathbf{a}) \equiv \left\| \frac{\partial \mathbf{b}}{\partial \mathbf{a}} \right\| \tag{4.21}$$

is the Jacobian determinant of the transformation $\mathbf{b} = \mathbf{F}(\mathbf{a})$. It follows from Eqs. (4.18)–(4.20) that the function $g(\mathbf{a})$ transforms in the same way; i.e.,

$$g(\mathbf{b}) = J^{-1}\, g(\mathbf{a}) \tag{4.22}$$

which of course reflects the fact that both $\bar{p}$ and g can be interpreted as volumetric densities of conserved quantities. Equation (4.22) states that once $g(\mathbf{a})$ has somehow been defined or determined for a particular choice of the variables $\mathbf{A}$, its functional dependence on $\mathbf{b} = \mathbf{F}(\mathbf{a})$ for an alternative equivalent set of variables $\mathbf{B} = \mathbf{F}(\mathbf{A})$ is uniquely determined and can no longer be tampered with. The Jacobian J depends on $\mathbf{a}$ or $\mathbf{b}$ for nonlinear tranformations, so if $g(\mathbf{a})$ is a constant independent of $\mathbf{a}$ then $g(\mathbf{b})$ must vary with $\mathbf{b}$ according to Eq. (4.22), and vice versa. It is therefore inconsistent to simply presume or require that $g(\mathbf{a})$ be constant in general, since this cannot be true for an arbitrary alternative equivalent choice of the variables $\mathbf{A}$. This situation is closely analogous to that discussed in connection with continuous state spaces in Sect. 2.6.

We do, however, have the freedom to define or require that $g(\mathbf{a}) = 1/v(\mathbf{a})$ be constant for one particular designated choice of the variables $\mathbf{A}$ (or linear combinations thereof, since J is constant for linear transformations). Just as in Sect. 2.6, such variables will be referred to as a *natural representation*, but in this case of the macrostates rather than the microstates of the system [23]. If the variables $\mathbf{a}$ constitute a natural representation then $g(\mathbf{a}) = g_a$, where g_a is a constant independent of $\mathbf{a}$, and the subscript a is again not a variable but merely a label which serves to identify the set of variables $\mathbf{a}$. It then follows that $M = \int d\mathbf{a}\, g(\mathbf{a}) = g_a V_a$, so that $\bar{v} = V_a/M = 1/g_a$ and $v(\mathbf{a}) = 1/g(\mathbf{a}) = 1/g_a = \bar{v}$ independently of $\mathbf{a}$. The local tolerance $v(\mathbf{a})$ associated with the constraint of Eq. (4.7) is then quantified by the constant parameter g_a, and no longer requires an entire function of $\mathbf{a}$ for its specification. The numerical value of g_a still remains arbitrary within the wide limits implied by Eq. (4.10), which now simply requires $g_a \gg 1/V_a$. As previously mentioned, however, the dependence of $S(\mathbf{a})$ on g_a is often so weak as to be negligible, as the development of Chapter 6 will illustrate.

The question then arises as to whether there exists a preferred set of variables $\mathbf{a}$ which merits such a distinction in any particular context, and if so how those variables are to be defined or determined. No general or absolute requirements or criteria exist for deciding which variables constitute a natural representation, but this ambiguity can be regarded as advantageous because it provides the freedom and opportunity to base that decision on utilitarian considerations such as convenience or simplicitly. However, a guiding principle has evolved to the effect that whenever possible, the variables in a natural representation should be *additive* in the same sense as the entropy itself. This principle is suggested by metric considerations that have not yet been discussed, so its motivation and elaboration will be deferred to the next subsection.

The most probable neighborhood $\mathbb{N}(\mathbf{a})$ in $\mathbf{a}$-space is that for which $W(\mathbf{a})/W$ is a maximum. The value of $\mathbf{a}$ at which that maximum occurs is presumed unique and is denoted by $\mathbf{a}^*$. Clearly $W(\mathbf{a})$ and $S(\mathbf{a}) = \log W(\mathbf{a})$ also attain their maximum values $W^* \equiv W(\mathbf{a}^*)$ and $S^* \equiv S(\mathbf{a}^*) = \log W^*$ at $\mathbf{a} = \mathbf{a}^*$, so that $\mathbf{a}^*$ is the macrostate of maximum entropy. It follows from Eq. (4.19) that $S(\mathbf{b})$ and $S(\mathbf{a})$ have the same maximum value $S^* = S(\mathbf{b}^*) = S(\mathbf{a}^*)$, where $\mathbf{b}^* = \mathbf{F}(\mathbf{a}^*)$, so the macrostate $\mathbf{a}^*$, like the entropy itself, is covariant. Equation

(4.18) shows that the maximum of $S(\mathbf{a})$ coincides with the maximum of $\bar{p}(\mathbf{a})/g(\mathbf{a})$, which implies that it does not in general coincide with the maximum of $\bar{p}(\mathbf{a})$ itself. However, the maxima of $S(\mathbf{a})$ and $\bar{p}(\mathbf{a})$ do coincide in the special case when the variables $\mathbf{A}$ are a natural representation, but only for those variables. In that case the maximum value of $\bar{p}(\mathbf{a})$ does occur at $\mathbf{a} = \mathbf{a}^*$, but the value $\mathbf{b}^* = \mathbf{F}(\mathbf{a}^*)$ then maximizes $\bar{p}(\mathbf{b})/g(\mathbf{b})$ rather than $\bar{p}(\mathbf{b})$. The macrostate which maximizes the probability density itself is therefore not covariant, which is also evident from Eq. (4.20).

Although $g(\mathbf{a})$ is essential for covariance, its functional form becomes essentially irrelevant in the important special case when there exists a subset $\circledast$ of a-space such that (i) $\bar{p}(\mathbf{a})$ is negligible for $\mathbf{a} \notin \circledast$, and (ii) $g(\mathbf{a})$ is nearly constant for $\mathbf{a} \in \circledast$. In practice, these conditions typically obtain only when W is very large and $\bar{p}(\mathbf{a})$ is sharply peaked about the point $\mathbf{a} = \mathbf{a}^*$. Condition (i) implies that $\mathbf{a}^* \in \circledast$ and

$$\int_{\circledast} d\mathbf{a}\, \bar{p}(\mathbf{a}) \cong 1 \tag{4.23}$$

whereupon condition (ii) implies that $g(\mathbf{a}) \cong g(\mathbf{a}^*)$ for $\mathbf{a} \in \circledast$. Equation (4.18) can then be approximated by

$$\bar{p}(\mathbf{a}) = g(\mathbf{a}^*) \exp\{S(\mathbf{a}) - S\} = \frac{\exp S(\mathbf{a})}{\int d\mathbf{a}' \exp S(\mathbf{a}')} \tag{4.24}$$

for $\mathbf{a} \in \circledast$. This approximation is inaccurate outside the region $\circledast$, but this is immaterial since $\bar{p}(\mathbf{a})$ is negligible there. Equation (4.24) has the same form that Eq. (4.18) would assume in a natural representation with $g(\mathbf{a}) = g(\mathbf{a}^*)$ for all $\mathbf{a}$. Thus we see that replacing $g(\mathbf{a})$ by a constant in Eq. (4.18), which would violate covariance in general, becomes permissible under conditions (i) and (ii) above, and in particular for probability densities which are sufficiently sharply peaked. Of course, the value of that constant is $g(\mathbf{a}^*)$, which depends upon the choice of the variables $\mathbf{A}$, as does the validity of condition (ii) and the functional form of $S(\mathbf{a})$.

Just as in Sect. 4.1, the question naturally arises as to the conditions under which the maximum value $S^* = S(\mathbf{a}^*)$ of the constrained entropy $S(\mathbf{a})$ provides an accurate approximation to the unconstrained entropy S. The fundamental condition for that approximation to be valid is again simply $\log W - \log W^* \ll \log W$, or

$$\log(W/W^*) \ll \log W \tag{4.25}$$

which is formally identical to Eq. (4.5). The discussion in Sect. 4.1 suggests that Eq. (4.25) is more likely to be satisfied when the probability density $\bar{p}(\mathbf{a})$ is sharply peaked or otherwise concentrated in a subset $\circledast$ of $\mathbf{a}$-space as described above, so we shall restrict attention to that case. Equations (4.14), (4.16), and (4.23) then combine to imply

$$W = \int_{\circledast} d\mathbf{a}\, g(\mathbf{a})\, W(\mathbf{a}) \tag{4.26}$$

which replaces Eq. (4.15) in the present context. Equation (4.26) simply reflects the fact that Eq. (4.23) implies that the microstates k are so highly concentrated in the region $\circledast$ that those lying outside that region are negligible. The squeeze technique described in Sect. 4.1 can now be applied to the integral in Eq. (4.26), whereby we obtain the obvious inequalities

$$\int_{\mathbb{N}^*} d\mathbf{a}\, g(\mathbf{a})\, W(\mathbf{a}) \;\le\; W \;\le\; M_{\circledast}\, W^* \tag{4.27}$$

where $\mathbb{N}^* \equiv \mathbb{N}(\mathbf{a}^*)$ is the neighborhood of the point $\mathbf{a}^*$, and $M_{\circledast} \equiv \int_{\circledast} d\mathbf{a}\, g(\mathbf{a})$ is the number of disjoint neighborhoods $\mathbb{N}(\mathbf{a})$ in the region $\circledast$. Thus $1/M_{\circledast}$ provides a covariant measure of how sharply peaked or concentrated the probability density is. Both $g(\mathbf{a})$ and $W(\mathbf{a})$ are nearly constant for $\mathbf{a} \in \mathbb{N}(\mathbf{a}^*)$, so they can be evaluated at $\mathbf{a} = \mathbf{a}^*$ and taken outside the integral in Eq. (4.27). We thereby obtain

$$W^* \;\le\; W \;\le\; M_{\circledast}\, W^* \tag{4.28}$$

where use has been made of Eqs. (4.11) and (4.16). Equation (4.28) is equivalent to

$$S^* \leq S \leq S^* + \log M_{\circledast} \tag{4.29}$$

which shows that $S \cong S^*$ if $\log M_{\circledast} \ll S = \log W$. That is to say, if the region $\circledast$ is small enough that $\log M_{\circledast} \ll \log W$, then the approximation $S \cong S^*$ is accurate. Note that the condition $M_{\circledast} \ll W$, which is trivially implied by Eqs. (4.9) and (4.26), is *not* by any means sufficient to imply $S \cong S^*$.

It is often useful and convenient to approximate sharply peaked probability distributions by multivariate Gaussian distributions. Such approximations are obtained simply by expanding $S(\mathrm{a})$ in a Taylor series about the point $\mathrm{a} = \mathrm{a}^*$ and truncating the series at the quadratic term. This yields

$$S(\mathrm{a}) \cong S^* - \tfrac{1}{2}\,(\mathrm{a} - \mathrm{a}^*) \cdot \mathbf{D} \cdot (\mathrm{a} - \mathrm{a}^*) \tag{4.30}$$

where $\mathbf{D} \equiv -\,(\partial^2 S/\partial \mathrm{a} \partial \mathrm{a})^*$, and the superscript $*$ implies evaluation at $\mathrm{a} = \mathrm{a}^*$ as usual. The linear term which would otherwise appear vanishes because $S(\mathrm{a})$ is a maximum at $\mathrm{a} = \mathrm{a}^*$, which implies that $(\partial S/\partial \mathrm{a})^* = 0$ and that $\mathbf{D}$ is positive definite. The Gaussian approximation to $\bar{p}(\mathrm{a})$ is then obtained by combining Eqs. (4.24) and (4.30), with the result

$$\bar{p}(\mathrm{a}) \cong \sqrt{\|\mathbf{D}\|/(2\pi)^d}\, \exp\{-\tfrac{1}{2}\,(\mathrm{a} - \mathrm{a}^*) \cdot \mathbf{D} \cdot (\mathrm{a} - \mathrm{a}^*)\} \tag{4.31}$$

where $\|\mathbf{D}\|$ is the determinant of $\mathbf{D}$ and d is the dimensionality of a. The approximate probability density of Eq. (4.31) correctly reproduces the second moments of $\bar{p}(\mathrm{a})$, but not in general the higher moments [29]. It is therefore useful in situations where it is sufficient to represent the overall width of the distribution reasonably well, but finer details are not of interest.

The accuracy considerations discussed in the last paragraph of Sect. 4.1 are equally applicable, *mutatis mutandis*, in the present context. Once again, the essential point is that accuracy of $S(\mathrm{a}) = \log W(\mathrm{a})$ merely requires logarithmic accuracy of $W(\mathrm{a})$, while relatively small errors in $S(\mathrm{a})$ are conversely magnified into relatively large errors in $W(\mathbf{a}) = \exp S(\mathbf{a})$. However, Eq. (4.18) shows that such errors tend to cancel out in $\bar{p}(\mathbf{a})$ due to the normalization. Moreover, the cancellation is nearly complete when $\bar{p}(\mathbf{a})$ is sharply peaked or otherwise concentrated in a region $\circledast$ and $\exp S(\mathbf{a})$ is in error by a nearly constant factor for $\mathbf{a} \in \circledast$, regardless of how large that factor may be. The smaller $\circledast$ and $M_{\circledast}$ are, the more likely that situation is to obtain, and the less likely small errors in $S(\mathbf{a})$ are to produce significant errors in $\bar{p}(\mathbf{a})$ as determined by Eq. (4.24). However, no general statement can be made as to how small $M_{\circledast}$ must be for the resulting errors in $\bar{p}(\mathbf{a})$ to be negligible, which clearly depends upon how the errors in $S(\mathbf{a})$ vary with $\mathbf{a}$, *inter alia.*

4.2.3 Metric Considerations

Covariance is normally discussed in connection with tensor analysis and differential geometry, which in their modern coordinate-free forms have evolved into elegant and powerful tools with many physical applications [30, 31]. The basic geometrical object in this formalism is the metric tensor, in terms of which geometrical concepts such as length, area, volume, angle, projection, etc. can be defined and quantified in a manner which is manifestly independent of the particular coordinates or variables chosen. Unfortunately, this elegance comes with a price: the formalism is subtle, abstract, and intricate, and its assimilation requires a significant intellectual investment comparable to learning a new language, which indeed it is. Moreover, like other power tools its use entails a certain risk of self-injury, which in this context is insidious because those injured may not be aware of it. Ironically, the very features of the formalism which make it advantageous and economical may also tend to obscure the essential simplicity of the underlying ideas. In many situations, however, the full machinery of the formalism is not required and its use would constitute excessive force. In particular, a full specification of the metric tensor is not always necessary. For present purposes it suffices to define the covariant volume element in state space, which depends only on the determinant of the metric tensor.

A covariant volume element $d\Gamma$ in state space can be defined with respect to a particular set of variables $\mathbf{A}$ by

$$d\Gamma = \gamma(\mathbf{a})\, d\mathbf{a} \tag{4.32}$$

where $\gamma(\mathbf{a})$ is a smooth positive function of $\mathbf{a}$. In metric terms $\gamma(\mathbf{a})$ is the square root of the determinant of the metric tensor [23, 30]. Thus defining $\gamma(\mathbf{a})$ imposes a restriction on the metric tensor but does not completely determine it, nor is there any need to do so, since the only metric concept needed in the present context is volume. The covariant volume elements $\gamma(\mathbf{b})\, d\mathbf{b}$ and $\gamma(\mathbf{a})\, d\mathbf{a}$ are equal by definition, so

$$\gamma(\mathbf{b}) = J^{-1}\, \gamma(\mathbf{a}) \tag{4.33}$$

where $J(\mathbf{a}) \equiv \|\partial\mathbf{b}/\partial\mathbf{a}\|$ as before. Equations (4.22) and (4.33) combine to imply

$$\frac{\gamma(\mathbf{b})}{g(\mathbf{b})} = \frac{\gamma(\mathbf{a})}{g(\mathbf{a})} \tag{4.34}$$

The covariant volume of any region $\mathbb{R}$ of $\mathbf{a}$-space is then given by

$$\Gamma_{\mathbb{R}} = \int_{\mathbb{R}} d\mathbf{a}\, \gamma(\mathbf{a}) = \int_{\mathbb{R}} d\mathbf{b}\, \gamma(\mathbf{b}) \tag{4.35}$$

The variables $\mathbf{a}$ and $\mathbf{b}$ will generally have different units, and there are no preferred variables or units associated with covariant volume, so $\Gamma_{\mathbb{R}}$ must be considered dimensionless. Thus $\gamma(\mathbf{a})$ has the inverse units of $d\mathbf{a}$. There is no unique, intrinsic, or *a priori* preferred metric tensor in $\mathbf{a}$-space, so neither is there a preferred volume element. Thus we are free to define $\gamma(\mathbf{a})$ at our discretion and convenience for some particular choice of the variables $\mathbf{a}$, whereupon $\gamma(\mathbf{b})$ is determined for any other set of variables $\mathbf{b} = \mathbf{F}(\mathbf{a})$ by Eq. (4.33).

We shall require $\gamma(\mathbf{a})$ to be a slowly varying function of $\mathbf{a}$ which is very nearly constant within any of the neighborhoods $\mathbb{N}(\mathbf{a})$, so that $\gamma(\mathbf{a}')$ can be accurately approximated by $\gamma(\mathbf{a})$ for $\mathbf{a}' \in \mathbb{N}(\mathbf{a})$. The covariant volume of the neighborhood $\mathbb{N}(\mathbf{a})$ is then given by

$$\Gamma(\mathbf{a}) \equiv \int_{\mathbb{N}(\mathbf{a})} d\mathbf{a}' \, \gamma(\mathbf{a}') \cong \gamma(\mathbf{a}) \, v(\mathbf{a}) = \frac{\gamma(\mathbf{a})}{g(\mathbf{a})} \tag{4.36}$$

the covariance of which is confirmed by Eq. (4.34) or (4.35). Since the neighborhood $\mathbb{N}(\mathbf{a})$ defines the tolerance associated with the constraint of Eq. (4.7), $\Gamma(\mathbf{a})$ can be interpreted as a covariant measure of that tolerance. (Of course, that tolerance also determines, or conversely is determined by, the function $g(\mathbf{a}) = 1/v(\mathbf{a})$ via Eq. (4.11). However, the value of $g(\mathbf{a})$ depends upon the particular variables $\mathbf{A}$ chosen to represent the macrostate of the system, whereas the value of $\Gamma(\mathbf{a})$ does not.) For reasons that will become clear as we proceed, it is natural, convenient, and advantageous to simply define

$$\gamma(\mathbf{a}) \equiv g(\mathbf{a}) \tag{4.37}$$

whereupon $\gamma(\mathbf{a})$ and $g(\mathbf{a})$ become synonymous. Equation (4.34) shows that this definition is in fact already covariant, so it applies for any choice of the variables $\mathbf{A}$. Combining Eqs. (4.36) and (4.37), we obtain

$$\Gamma(\mathbf{a}) = 1 \tag{4.38}$$

Thus the definition of Eq. (4.37) has the felicitous consequence that when the tolerance of the constraint $\mathbf{A}_k \in \mathbb{N}(\mathbf{a})$ is quantified by the *covariant* volume $\Gamma(\mathbf{a})$ of the neighborhood $\mathbb{N}(\mathbf{a})$, it becomes a constant independent of $\mathbf{a}$ which is the same for all neighborhoods $\mathbb{N}(\mathbf{a})$ and for any choice of the variables $\mathbf{A}$. This obviously represents a significant simplification. According to Eq. (4.38), the numerical value of that constant covariant tolerance is unity, which does not superficially appear to be small as it presumably should be. However, its size must be interpreted relative to the total covariant volume of a-space, which is given by $\Gamma = \int d\mathbf{a} \, \gamma(\mathbf{a}) = \int d\mathbf{a} \, g(\mathbf{a}) = M$, where $M \gg 1$ is the total number of disjoint neighborhoods in a-space as discussed in Sect. 4.2.1. Thus $\Gamma(\mathbf{a})/\Gamma = 1/M \ll 1$, which is the same as the dimensionless mean or global tolerance introduced in Sect. 4.2.1.

However, the main virtue of Eq. (4.37) does not reside in its tolerance implications, but rather in the fact that it combines with Eq. (4.32) to imply

$$d\Gamma = g(\mathbf{a})\,d\mathbf{a} \tag{4.39}$$

which invests $g(\mathbf{a})$ with a metric interpretation in terms of the covariant volume element in state space. This new interpretation serves as a guide as to which choice of the variables $\mathbf{a}$ should be considered a natural representation in which $g(\mathbf{a})$ is a constant independent of $\mathbf{a}$. For such variables, the volume element $d\Gamma$ differs from $d\mathbf{a}$ only by a constant factor. But $d\mathbf{a} = \prod_i da_i$ is simply the standard expression for the volume element in a Euclidean space in terms of Cartesian coordinates $(a_1, a_2, \cdots)$. This suggests that the variables $\mathbf{A}$ in a natural representation can be interpreted as Cartesian coordinates in an abstract macroscopic state space which is *globally* Euclidean (and not merely locally, which of course is always the case). The significance of this interpretation is that such a space is moreover a *linear* space in which Cartesian coordinates are *additive*; i.e., the equation $\mathbf{A} = \mathbf{A}_1 + \mathbf{A}_2$ is meaningful and sensible, and the resulting sum $\mathbf{A}$ has the same physical and/or mathematical significance as the separate terms $\mathbf{A}_1$ and $\mathbf{A}_2$. Additivity will be discussed in more detail momentarily, but the above observations already suggest that the variables in a natural representation should generally be expected or required to be additive. Conversely, the property of additivity thereby suggests itself as a natural criterion which the variables $\mathbf{A}$ should satisfy to be considered a natural representation. Indeed, at present it is the only such criterion at our disposal, and in its absence the choice of a natural representation would remain arbitrary and ambiguous.

Additive variables typically represent the quantity or amount of some measurable property, for which a corresponding conservation law often exists. Such variables as length, area, volume, mass, energy, magnetization, etc. are therefore additive, whereas nonlinear functions of such variables (e.g., the cube root of area or the reciprocal of energy) do not have the same significance and are not additive. Thus mere dimensional consistency does not imply additivity. Extensive thermodynamic variables are additive in the sense described above, while intensive variables are not, but additivity is a more general and fundamental property than extensivity.

It may not always be obvious whether or not a set of additive variables $\mathbf{A}$ even exists. If it does then a-space is isomorphic to a linear space, albeit possibly in disguise if the natural representation is not recognized and the macrostate of the system is specified by some other variables. For example, if the points in a Euclidean plane are initially parameterized in terms of the polar coordinates (r, θ), which of course are not additive, it is not immediately obvious that they can be transformed into the additive rectangular Cartesian coordinates $(x, y) = (r \cos \theta, r \sin \theta)$. In situations of physical interest, however, the additive variables that constitute a natural representation are usually fairly obvious. Foremost amongst those variables, and by far the most important in physical applications, is the energy.

The above considerations strongly suggest the guiding principle that whenever possible, the variables $\mathbf{A}$ which constitute a natural representation should be chosen to be additive in the sense outlined above. The property of additivity is particularly pertinent to systems which are, or can be regarded as, composed or made up of two or more parts or subsystems, each of which has its own microstates with well defined values of the macrostate variables $\mathbf{A}$. In particular, systems of this type arise when two separate systems $\mathcal{A}$ and $\mathcal{B}$ with the same macrostate variables $\mathbf{A}$ are regarded as a single combined or composite system $\mathcal{AB}$. Let the microstates of two such systems $\mathcal{A}$ and $\mathcal{B}$ be labeled by the indices i and j respectively, and denote the corresponding values of the additive variables $\mathbf{A}$ by $\mathbf{A}_i^{\mathcal{A}}$ and $\mathbf{A}_j^{\mathcal{B}}$. The microstates of the combined system $\mathcal{AB}$ are then labeled by the ordered pairs $k = (i, j)$, and the value of the additive quantity $\mathbf{A}$ when system $\mathcal{AB}$ occupies microstate $k = (i, j)$ is given by

$$\mathbf{A}_k^{\mathcal{AB}} = \mathbf{A}_{ij}^{\mathcal{AB}} = \mathbf{A}_i^{\mathcal{A}} + \mathbf{A}_j^{\mathcal{B}} \tag{4.40}$$

Note that this implies additivity of the corresponding macrostates $\mathbf{a}$ as well: if $\mathbf{A}_i^{\mathcal{A}} \cong \mathbf{a}_{\mathcal{A}}$ and $\mathbf{A}_j^{\mathcal{B}} \cong \mathbf{a}_{\mathcal{B}}$ then $\mathbf{A}_i^{\mathcal{A}} + \mathbf{A}_j^{\mathcal{B}} \cong \mathbf{a}_{\mathcal{A}} + \mathbf{a}_{\mathcal{B}}$.

The composite systems $\mathcal{AB}$ of greatest interest are those in which the subsystems $\mathcal{A}$ and $\mathcal{B}$ interact with each other and therefore cannot be presumed to be statistically independent. In such situations it is often of interest to focus attention on one of the subsystems, say system $\mathcal{A}$, and analyze the effects of imposing and relaxing constraints of the form of Eq. (4.7) only on the variables $\mathbf{A}_i^{\mathcal{A}}$. The general formalism developed in this section can be applied to such situations simply by defining $\mathbf{A}_k = \mathbf{A}_{ij} = \mathbf{A}_i^{\mathcal{A}}$ for all j, so that $\mathbf{A}_k$ represents the values of the macrostate variables $\mathbf{A}$ for system $\mathcal{A}$ alone. Of course, the microstates j of system $\mathcal{B}$ nevertheless remain present and must be accounted for in all sums over the microstates $k = (i, j)$. Equation (4.7) and the remainder of the formalism are then formally unchanged, but it must be remembered that the macrostate $\mathbf{a}$ now represents the constrained value of $\mathbf{A}_i^{\mathcal{A}}$ alone rather than that of $\mathbf{A}_k^{\mathcal{AB}}$, and hence does not include the values of the quantities $\mathbf{A}_j^{\mathcal{B}}$ associated with system $\mathcal{B}$. As discussed at the beginning of this chapter, it must also be remembered that in general the microstates $k = (i, j)$ are implicitly subject to further parameterized constraints to ensure that the total number of accessible microstates W is finite, the most common of which are constraints of the form of Eq. (4.7) on some or all of the variables $\mathbf{A}_k^{\mathcal{AB}}$. These admittedly somewhat abstract considerations will be clarified and exemplified by the treatment of macroscopic thermodynamic systems in Chapter 6.

We remark parenthetically that it is not sensible to speak of the total amount of an additive quantity $\mathbf{A}$ in the group of microstates within a neighborhood $\mathbb{N}(\mathbf{a})$, or for that matter any other group of microstates. A system only occupies one of its microstates at any given time. As the above discussion illustrates, however, it is sensible to consider the total amount of an additive quantity $\mathbf{A}$ possessed by two or more systems or subsystems, each of which occupies its own microstate.

The formulation in this section has been developed for discrete microstates k. The corresponding development for continuous microstates $\mathbf{x}$ follows the same outline, so the details will be omitted, but it is well to indicate the principal modifications required. The macrostate variables $\mathbf{A}$ now become functions of the continuous variables $\mathbf{x}$ instead of the discrete index k, so $\mathbf{A}(\mathbf{x})$ replaces $\mathbf{A}_k$ in the constraint of Eq. (4.7). That constraint then implicitly defines the region $\mathbb{A}$ of $\mathbf{x}$-space which corresponds to the neighborhood $\mathbb{N}(\mathbf{a})$ in $\mathbf{a}$-space. The number of microstates consistent with that constraint is then given by $W(\mathbf{a}) = \int_{\mathbb{A}} d\mathbf{x}\, \rho(\mathbf{x})$, where $\rho(\mathbf{x})$ is the density of microstates in $\mathbf{x}$-space introduced in Sect. 2.6. The entropy $S(\mathbf{a})$ of the constrained system is then given by Eq. (4.8) as before. As discussed in Sect. 2.6, the probability density in $\mathbf{x}$-space for equally probable continuous microstates in the unconstrained system is simply $p_0(\mathbf{x}) = (1/W)\,\rho(\mathbf{x})$, where $W = \int d\mathbf{x}\, \rho(\mathbf{x})$ is the total number of microstates in the system. The average value of an arbitrary function $F(\mathbf{A})$ of the state variables $\mathbf{A}(\mathbf{x})$ in the unconstrained system is therefore given by

$$\langle F(\mathbf{A})\rangle = \int d\mathbf{x}\, p_0(\mathbf{x})\, F(\mathbf{A}(\mathbf{x})) = \int d\mathbf{a}\, p(\mathbf{a})\, F(\mathbf{a}) \qquad (4.41)$$

where

$$p(\mathbf{a}) = \int d\mathbf{x}\, p_0(\mathbf{x})\, \delta(\mathbf{a} - \mathbf{A}(\mathbf{x})) = \langle \delta(\mathbf{a} - \mathbf{A}(\mathbf{x}))\rangle \qquad (4.42)$$

is the true probability density in $\mathbf{a}$-space in the unconstrained system. In contrast to the case of discrete microstates, however, $p(\mathbf{a})$ is now already continuous, so there is no longer any need or motivation to introduce a smoothed mean probability density $\bar{p}(\mathbf{a})$ as was done for discrete microstates in Eqs. (4.12)–(4.14). The remainder of the development is then essentially indifferent to whether the microstates are discrete or continuous. In particular, the fundamental Eqs. (4.17) and (4.18) remain valid for continuous microstates $\mathbf{x}$ with $\bar{p}(\mathbf{a})$ replaced by $p(\mathbf{a})$ as given by Eq. (4.42). It warrants reemphasis that $p(\mathbf{a})$ and $\bar{p}(\mathbf{a})$ themselves are entirely independent of and unrelated to the neighborhoods $\mathbb{N}(\mathbf{a})$ and their volumes $v(\mathbf{a}) = 1/g(\mathbf{a})$. However, the neighborhoods must be introduced in order to properly define $W(\mathbf{a})$ and $S(\mathbf{a})$, which is why the relations between $p(\mathbf{a})$ or $\bar{p}(\mathbf{a})$ and $W(\mathbf{a})$ or $S(\mathbf{a})$ necessarily involve $v(\mathbf{a})$ or $g(\mathbf{a})$ as well.

4.2.4 Mean Constraints

In contrast to sharp constraints, which render some (in fact usually most) of the equally probable microstates k strictly inaccessible while leaving the remainder equally probable, we define mean constraints as conditions imposed on mean or average values which thereby require the probabilities p_k to become unequal, so that some of the microstates become more probable at the expense of others which become less probable. The resulting unequal probabilities then imply that the entropy is no longer the BP entropy $S = \log W$ but must now be evaluated as the BGS entropy $S = -\sum_k p_k \log p_k$. Since the probabilities p_k must sum to unity, it is clear that the maximum possible number of mean constraints is $W - 1$, in which case the probabilities are uniquely determined (provided the constraints are independent and consistent with one another and with the condition $0 \leq p_k \leq 1$). Normally, however, the number of mean constraints is very much smaller than the number of states W, so that the probabilities p_k, and thereby the entropy as well, are left underdetermined unless or until further conditions are imposed.

A set of mean constraints on a set of observables $\mathbf{A} = (A_1, A_2, \cdots)$ takes the general form

$$\sum_k p_k \mathbf{A}_k = \bar{\mathbf{a}} \tag{4.43}$$

for systems with discrete states, or

$$\int d\mathbf{x}\, p(\mathbf{x})\, \mathbf{A}(\mathbf{x}) = \bar{\mathbf{a}} \tag{4.44}$$

for systems with continuous states, where $\bar{\mathbf{a}}$ is the specified mean value of $\mathbf{A}_k$ or $\mathbf{A}(\mathbf{x})$. The remainder of the discussion in this chapter will be restricted to discrete states for brevity, but the analogous considerations for continuous states are entirely straightforward.

Since mean constraints do not require a tolerance to be specified, they are more precisely defined and more convenient to work with than sharp constraints. Moreover, they become essentially equivalent to sharp constraints in situations where the probability density of the variables $\mathbf{A}_k$ is sharply peaked about its most probable value. Unlike sharp constraints, however, mean constraints are not covariant with respect to nonlinear transformations of the state variables $\mathbf{A}_k$, because if $\mathbf{B}_k = \mathbf{F}(\mathbf{A}_k)$ and $\bar{\mathbf{b}} = \mathbf{F}(\bar{\mathbf{a}})$ then $\sum_k p_k \mathbf{B}_k \neq \bar{\mathbf{b}}$ unless the function $\mathbf{F}$ is linear. (Alternatively, if $\sum_k p_k \mathbf{B}_k = \bar{\mathbf{b}}$ then $\bar{\mathbf{b}} \neq \mathbf{F}(\bar{\mathbf{a}})$.) However, mean constraints are covariant with respect to linear transformations of the state variables, which in general are sensible only for additive variables.

Note that additional mean constraints on arbitrary nonlinear functions of the observables $\mathbf{A}_k$ can be imposed within the same framework simply by enlarging the set of variables to include those functions. For example, suppose we wish to constrain both the mean value and the variance of a single observable A_k to have the specified values $\sum_k p_k A_k = \bar{a}$ and $\sum_k p_k (A_k - \bar{a})^2 = \delta a^2$, respectively. The additional constraint on the variance is equivalent to $\sum_k A_k^2 = \bar{a}^2 + \delta a^2$, so both constraints can be simultaneously imposed simply by defining $\mathbf{A}_k = (A_k, A_k^2)$ and $\bar{\mathbf{a}} = (\bar{a}, \bar{a}^2 + \delta a^2)$.

Equation (4.43) shows that mean constraints can be regarded and interpreted as parameterized linear constraints on the values of the probabilities p_k, whereas the sharp constraints of Eq. (4.7) are parameterized constraints on the states k themselves. Since the number of components of $\mathbf{A}_k$ is ordinarily much smaller than W, Eq. (4.43) is ordinarily insufficient to uniquely determine either the probabilities p_k or the BGS entropy $S(\bar{\mathbf{a}})$ of the constrained system. Further conditions must therefore be imposed to obtain definite values of p_k and $S(\bar{\mathbf{a}})$. Those conditions are normally obtained from the principle of maximum entropy, as discussed in the next chapter. Nevertheless, it is clear that Eq. (4.43) alone already implies that the probabilities p_k cannot be equal (unless $W\bar{\mathbf{a}} = \sum_k \mathbf{A}_k$, which is an insignificant degenerate case). It follows that $S(\bar{\mathbf{a}}) < \log W$, since the BGS entropy attains its maximum value of $\log W$ when the probabilities are equal with the common value $p_k = 1/W$. Mean constraints therefore reduce the entropy of the system, albeit by an as yet indeterminate amount. Conversely, when such constraints are removed the entropy cannot decrease and will ordinarily increase back to its previous unconstrained value of $\log W$, provided that some mechanism exists for producing spontaneous transitions between the microstates k as discussed in Sect. 4.1.

4.2.5 Mixed Constraints

It is straightforward to combine the preceding ingredients to obtain a description of systems subjected to mixed constraints; i.e., systems which are simultaneously subjected to sharp constraints on certain state variables and mean constraints on others. Such situations can easily be treated by imposing the sharp constraints first, which renders some (again, normally most) of the microstates k strictly inaccessible. Those microstates then have vanishing probabilities and can be omitted from consideration, whereupon the mean constraints can then be imposed on the microstates which remain accessible. If the state variables subjected to sharp constraints are denoted by $\mathbf{A}_k^s$, then Eq. (4.7) remains in effect for those variables; i.e.,

$$\mathbf{A}_k^s \in \mathbb{N}(\mathbf{a}) \tag{4.45}$$

The set of microstates k consistent with Eq. (4.45) will be denoted by $\mathbb{K}(\mathbf{a})$. If the state variables subjected to mean constraints are denoted by $\mathbf{A}_k^m$, then Eq. (4.43) applies to those variables:

$$\sum_{k \in \mathbb{K}(\mathbf{a})} p_k \mathbf{A}_k^m = \bar{\mathbf{a}} \tag{4.46}$$

As before, Eq. (4.46) will not normally suffice to determine the probabilities p_k uniquely, so the principle of maximum entropy is again ordinarily invoked for that purpose. The resulting probabilities p_k will then clearly retain a parametric dependence on $\mathbf{a}$ as well as $\bar{\mathbf{a}}$.

Chapter 5

THE PRINCIPLE OF MAXIMUM ENTROPY

5.1 Motivation

A conspicuous aspect of everyday life is that there is a general tendency for uncertainty to increase with the passage of time, and that resisting this tendency requires a continual expenditure of effort. If we throw a bottle into the ocean or release a balloon into the atmosphere, its subsequent whereabouts rapidly become increasingly uncertain. If we don't put our tools back in the toolbox after using them, they get scattered around and we can't find them when we need them. If librarians didn't return books to their proper place on the shelves, any particular book could equally well be anywhere in the library. These and similar mundane examples can all be interpreted as a tendency for the location of an object to become increasingly uncertain; i.e., an increase in the number of possible locations where the object might be found. In more general terms, this corresponds to an increase in the number of states which the system might possibly occupy. These examples also share the common feature that there is some process or mechanism at work which produces transitions between the accessible locations or states, as discussed in Sect. 4.1. In situations where the uncertainty is bounded (e.g., if a book is known to be somewhere in the library), its continual increase will inexorably result in a situation of maximum uncertainty in which no further increase is possible.

Since entropy is intended to be a quantitative measure of uncertainty, the mathematics of entropy as developed in the preceding chapters would be expected to reflect the general tendency for uncertainty to increase. And indeed it does, as the following observations attest:

(1) As shown in Sects. 2.5 and 2.6, the BGS entropy S attains its maximum value when the states of a system are equally probable, which intuitively represents a situation of maximum uncertainty. This implies that if S is regarded as a function of the probabilities p_k and is maximized with respect to all possible nonnegative values thereof (subject only to the normalization condition $\sum_k p_k = 1$), the result is $p_k = 1/W$. That is to say, equal probabilities imply maximum entropy, and conversely maximum entropy implies equal probabilities.

(2) As discussed in Sects. 4.1 and 4.2.4, the entropy cannot increase and will ordinarily decrease when either sharp or mean constraints are imposed, and conversely cannot decrease and will ordinarily increase when such constraints are removed. The resulting increase in entropy represents the increase in the uncertainty as to which state the system occupies which occurs when previously inaccessible states are made accessible, or when previously unequal probabilities are allowed to become equal.

Note that observation (1) above applies equally well in situations where some of the states of the system are inaccessible while the remaining accessible states are known or presumed to be equally probable. As discussed in Chapters 3 and 4, such situations occur when (a) the EAPP hypothesis is invoked and/or (b) sharp constraints are imposed on the system. In both cases, the equal probabilities of the accessible states imply that the entropy of the system has its maximum possible value subject to the restriction that the remaining states are inaccessible. Thus there is an intimate reciprocal relationship between the EAPP hypothesis and the principle of maximum entropy.

Up to this point we have simply regarded the BGS entropy S as a function or functional of the probabilities p_k or $p(\mathbf{x})$. Observations (1) and (2) above suggest that it may conversely be useful to regard the probabilities as being determined by the entropy, or more precisely by requiring the entropy or uncertainty to be as large as possible subject to whatever constraints may be in effect. Of course, those constraints always include the normalization condition $\sum_k p_k = 1$ for discrete states or $\int d\mathbf{x}\, p(\mathbf{x}) = 1$ for continuous states, but in general there will be other constraints as well. As noted in (1) above, these two converse views of the relation between entropy and probabilities are equivalent for sharp constraints. In contrast, however, mean constraints normally leave the probabilities, and hence the entropy as well, underdetermined as discussed in Sect. 4.2.4. Since the BGS entropy is a maximum for sharp constraints, it seems natural to require it to be a maximum for mean constraints as well. As will be seen in the next section, this requirement determines the probabilities (which in general are then unequal) and thereby removes the indeterminacy associated with mean constraints. The constrained maximization of the BGS entropy therefore constitutes a general procedure that unifies the treatment of both types of constraints. This intuitively appealing procedure is remarkable for its simplicity, generality, and a very wide variety of successful applications which lie outside the scope of the present discussion. It therefore seems appropriate to elevate it to the status of a principle, which is usually referred to as the Principle of Maximum Entropy (PME) [20, 21].

The PME now quantifies the reduction or increase in the BGS entropy that occurs when a mean constraint is imposed or relaxed, which was previously indeterminate as discussed in Sect. 4.2.4. Those changes can now be interpreted as trivial consequences of the obvious fact that a constrained maximum cannot exceed an unconstrained maximum. More generally, the PME implies that imposing an additional mean constraint cannot increase the entropy, and will normally further reduce it, since it excludes from consideration certain possible combinations of values of the probabilities p_k or $p(\mathbf{x})$ for which the previous maximum might have occurred.

The PME has been attributed to Jaynes, who certainly did more than anyone else to popularize and promulgate it, but its roots go back to Boltzmann and Gibbs. It is analogous to, although somewhat different in character from, the fundamental variational principles of physics, such as the principle of least action. Such principles typically originate as observed implications of more basic or primitive formulations, to which they are at first seen to be postulationally equivalent. Upon further scrutiny, however, they are ultimately perceived as more fundamental and fruitful than the original formulations from whence they sprung. So it is with the PME, which exhibits those very features.

Aside from the terminology, the present approach to entropy as a quantitative measure of uncertainty is entirely equivalent to the information theory approach [11, 20, 21, 32–35], in which the BGS entropy is referred to as information entropy and is interpreted as a measure of missing information; i.e., the information which would need to be acquired in order to determine the state of the system with complete certainty. This equivalence simply reflects the intuitive notion that information is the complement of uncertainty. In the language of information theory, the rationale for the PME is that it determines the probability distribution which contains the minimum information consistent with the constraints. It is therefore the least biased, most objective, and most noncommittal of all the possible probability distributions that satisfy the constraints. Conversely, any other probability distribution implies additional information which we do not in fact possess, and which should therefore not be allowed to influence our deliberations or calculations.

5.2 Generalized Canonical Probability Distributions

We now proceed to invoke the PME as a general procedure for determining discrete probability distributions p_k and continuous probability densities $p(\mathbf{x})$ for systems subject to mean constraints.

5.2.1 Discrete States

In systems with discrete states, mean constraints take the form of Eq. (4.43):

$$\langle \mathbf{A}_k \rangle \equiv \sum_k p_k \mathbf{A}_k = \bar{\mathbf{a}} \tag{5.1}$$

where $\bar{\mathbf{a}}$ is the specified mean value of $\mathbf{A}_k$. Of course, the probabilities p_k must also satisfy the usual normalization condition

$$\sum_k p_k = 1 \tag{5.2}$$

According to the PME, the probabilities p_k are to be determined by maximizing the BGS entropy S of Eq. (2.10) subject to the constraints of Eqs. (5.1) and (5.2). This can be accomplished by means of Lagrange's method of undetermined multipliers, with which the reader is presumed to be familiar. In order to apply this method in the present context, we must formally compute the *un*constrained maximum of the quantity

$$S_\alpha \equiv - \sum_k p_k \log p_k + \alpha_0 \sum_k p_k + \boldsymbol{\alpha} \cdot \sum_k p_k \mathbf{A}_k \tag{5.3}$$

with respect to the probabilities p_k, where α_0 and $\boldsymbol{\alpha} = (\alpha_1, \alpha_2, \cdots)$ are parameters which are implicitly determined *ex post facto* by enforcing Eqs. (5.1) and (5.2). Setting $\partial S_\alpha / \partial p_k = 0$, we obtain

$$\log p_k = \alpha_0 - 1 + \boldsymbol{\alpha} \cdot \mathbf{A}_k \tag{5.4}$$

so that the probability distribution has the form

$$p_k = (1/Z) \exp \boldsymbol{\alpha} \cdot \mathbf{A}_k \tag{5.5}$$

where $Z \equiv \exp(1 - \alpha_0)$. Equation (5.5) defines the generalized canonical probability distribution for discrete states. Equations (5.1), (5.2), and (5.5) combine to imply that

$$Z = Z(\boldsymbol{\alpha}) = \sum_k \exp \boldsymbol{\alpha} \cdot \mathbf{A}_k \tag{5.6}$$

and

$$\bar{\mathbf{a}} = \frac{1}{Z} \sum_k \mathbf{A}_k \exp \boldsymbol{\alpha} \cdot \mathbf{A}_k \tag{5.7}$$

The quantity Z is called the *partition function*. Equations (5.6) and (5.7) combine to express $\bar{\mathbf{a}}$ as an explicit function of $\boldsymbol{\alpha}$, which implicitly determines $\boldsymbol{\alpha}$ as a function of $\bar{\mathbf{a}}$. The probabilities p_k thereby depend parametrically on the value of $\bar{\mathbf{a}}$, as of course they must. When it is desirable to emphasize this dependence, p_k will be written as $p_k(\bar{\mathbf{a}})$. For historical reasons, the probability distribution $p_k(\bar{\mathbf{a}})$ given by Eq. (5.5) will be referred to as a generalized canonical probability distribution.

The probability distribution p_k can be used to compute the average of any desired state variables or functions thereof in the constrained system. It is of particular interest to evaluate the average of an arbitrary function $\mathbf{F}(\mathbf{A}_k)$ of the constrained variables $\mathbf{A}_k$ themselves, which is given by

$$\langle \mathbf{F}(\mathbf{A}_k) \rangle \equiv \sum_k p_k \, \mathbf{F}(\mathbf{A}_k) = \int d\mathbf{a} \, p(\mathbf{a}|\bar{\mathbf{a}}) \, \mathbf{F}(\mathbf{a}) \tag{5.8}$$

where

$$p(\mathbf{a}|\bar{\mathbf{a}}) \equiv \sum_k p_k(\bar{\mathbf{a}}) \, \delta(\mathbf{a} - \mathbf{A}_k) \tag{5.9}$$

is the true discontinuous probability density in a-space, which is clearly normalized so that $\int d\mathbf{a} \, p(\mathbf{a}|\bar{\mathbf{a}}) = 1$. We also note that $\int d\mathbf{a} \, p(\mathbf{a}|\bar{\mathbf{a}}) \, \mathbf{a} = \langle \mathbf{A}_k \rangle = \bar{\mathbf{a}}$. Combining Eqs. (5.5) and (5.9), we obtain

$$p(\mathbf{a}|\bar{\mathbf{a}}) = (1/Z) \, \omega(\mathbf{a}) \exp \boldsymbol{\alpha} \cdot \mathbf{a} \tag{5.10}$$

where

$$\omega(\mathbf{a}) \equiv \sum_k \delta(\mathbf{a} - \mathbf{A}_k) \tag{5.11}$$

is the true discontinuous density of states in **a**-space, which of course is an intrinsic property of the system, independent of $\bar{\mathbf{a}}$, $\boldsymbol{\alpha}$, or the probabilities p_k.

The partition function $Z(\boldsymbol{\alpha})$ might at first appear to be a mere normalization factor, but it is much more than that; it serves as a potential function from which various quantities of interest can be obtained by differentiation. The simplest such relation is obtained by differentiating Eq. (5.6) with respect to $\boldsymbol{\alpha}$ and combining the result with Eq. (5.7) to obtain

$$\bar{\mathbf{a}} = \frac{\partial \log Z}{\partial \boldsymbol{\alpha}} \tag{5.12}$$

which again explicitly expresses $\bar{\mathbf{a}}$ as a function of $\boldsymbol{\alpha}$, and conversely implicitly determines $\boldsymbol{\alpha}$ as a function of $\bar{\mathbf{a}}$. It is apparent that higher derivatives of Z with respect to $\boldsymbol{\alpha}$ are similarly related to averages of products of the state variables $\mathbf{A}_k$, and thereby to fluctuations in their values. For example,

$$\frac{1}{Z}\left(\frac{\partial^2 Z}{\partial \boldsymbol{\alpha}\, \partial \boldsymbol{\alpha}}\right) = \sum_k p_k\, \mathbf{A}_k \mathbf{A}_k = \langle \mathbf{A}_k \mathbf{A}_k \rangle = \int d\mathbf{a}\; p(\mathbf{a}|\bar{\mathbf{a}})\, \mathbf{a}\mathbf{a} \tag{5.13}$$

which is equivalent to

$$\frac{\partial^2 \log Z}{\partial \boldsymbol{\alpha}\, \partial \boldsymbol{\alpha}} = \langle (\mathbf{A}_k - \bar{\mathbf{a}})(\mathbf{A}_k - \bar{\mathbf{a}}) \rangle = \int d\mathbf{a}\; p(\mathbf{a}|\bar{\mathbf{a}})\, (\mathbf{a} - \bar{\mathbf{a}})(\mathbf{a} - \bar{\mathbf{a}}) \tag{5.14}$$

Thus $(\partial^2 \log Z / \partial \boldsymbol{\alpha}\, \partial \boldsymbol{\alpha})$ is simply the statistical covariance matrix of the variables $\mathbf{A}_k$.

As Eq. (5.13) illustrates, averages of products of $\mathbf{A}_k$ are simply moments of the probability density $p(\mathbf{a}|\bar{\mathbf{a}})$, and it is well known that knowledge of all its moments is normally sufficient to determine a probability density uniquely. Since all moments of $p(\mathbf{a}|\bar{\mathbf{a}})$ can evidently be expressed in terms of derivatives of $Z(\boldsymbol{\alpha})$, $Z(\boldsymbol{\alpha})$ consequently determines, at least in principle, not merely simple low-order averages such as $\langle \mathbf{A}_k \mathbf{A}_k \rangle$ but the entire probability density $p(\mathbf{a}|\bar{\mathbf{a}})$. This fundamental result deserves a less pedestrian derivation, and indeed is a straightforward consequence of the fact that $Z(\boldsymbol{\alpha})$ can be expressed in the alternative form

$$Z(\boldsymbol{\alpha}) = \int d\mathbf{a} \exp(\boldsymbol{\alpha}\cdot\mathbf{a})\, \omega(\mathbf{a}) \tag{5.15}$$

which follows immediately from Eq. (5.10), and reduces to Eq. (5.6) when combined with Eq. (5.11). Equation (5.15) shows that $Z(\boldsymbol{\alpha})$ is a multidimensional Laplace transform of the density of states $\omega(\mathbf{a})$, which reflects the fact that the Laplace transform of a sum of delta functions is a corresponding sum of exponentials. This transform could in principle be inverted to obtain $\omega(\mathbf{a})$, which would then immediately determine the probability density $p(\mathbf{a}|\bar{\mathbf{a}})$ via Eq. (5.10).

Now that the probability distribution p_k has been determined, the BGS entropy itself can be evaluated. Combining Eqs. (2.10) and (5.5), we obtain

$$S = \log Z - \boldsymbol{\alpha}\cdot\bar{\mathbf{a}} \tag{5.16}$$

where use has been made of Eqs. (5.6) and (5.7). Clearly S can be regarded as a function of either $\boldsymbol{\alpha}$ or $\bar{\mathbf{a}}$, although the latter is more natural. When it is desirable to emphasize this dependence, S will be written as $S(\bar{\mathbf{a}})$. Thus we have

$$S(\bar{\mathbf{a}}) = -\sum_k p_k(\bar{\mathbf{a}}) \log p_k(\bar{\mathbf{a}}) \tag{5.17}$$

Equation (5.12) shows that $d\log Z = \bar{\mathbf{a}} \cdot d\boldsymbol{\alpha}$, which combines with Eq. (5.16) to imply that $dS = -\boldsymbol{\alpha} \cdot d\bar{\mathbf{a}}$. Thus

$$\boldsymbol{\alpha} = -\frac{\partial S}{\partial \bar{\mathbf{a}}} \tag{5.18}$$

It then follows from Eqs. (5.12) and (5.18) that

$$\boldsymbol{\alpha} \cdot \bar{\mathbf{a}} \;=\; \boldsymbol{\alpha} \cdot \frac{\partial \log Z}{\partial \boldsymbol{\alpha}} \;=\; -\,\bar{\mathbf{a}} \cdot \frac{\partial S}{\partial \bar{\mathbf{a}}} \tag{5.19}$$

Equation (5.16) can therefore be rewritten in the equivalent forms

$$S \;=\; \log Z \;-\; \boldsymbol{\alpha} \cdot \frac{\partial \log Z}{\partial \boldsymbol{\alpha}} \;=\; \log Z \;+\; \bar{\mathbf{a}} \cdot \frac{\partial S}{\partial \bar{\mathbf{a}}} \tag{5.20}$$

It is well to confirm that this formulation properly reduces to the special case when there are no mean constraints, in which we know that the probability distribution and entropy are given by $p_k = 1/W$ and $S = \log W$, where W is the total number of equally probable discrete states. The mean constraints can be removed from the formalism simply by deleting Eq. (5.1) and setting $\boldsymbol{\alpha} = \mathbf{0}$, thereby leaving the normalization condition of Eq. (5.2) as the only remaining restriction on the probabilities. Equation (5.6) then reduces to $Z = W$, whereupon Eq. (5.5) reduces to $p_k = 1/W$ and Eq. (5.16) reduces to $S = \log W$.

5.2.2 Continuous States

The treatment of continuous states closely parallels the above development for discrete states, so some of the details will be omitted. Apart from the obvious replacement of sums by integrals, the only significant new feature in the continuous case is the density of states $\rho(\mathbf{x})$ discussed in Chapter 2. Most of the expressions for discrete states can be formally converted into the corresponding expressions for continuous states by means of the replacements

$$p_k \rightarrow \frac{p(\mathbf{x})}{\rho(\mathbf{x})} \quad ; \quad \sum_k \rightarrow \int d\mathbf{x}\, \rho(\mathbf{x}) \quad ; \quad \mathbf{A}_k \rightarrow \mathbf{A}(\mathbf{x}) \tag{5.21}$$

Thus the mean constraints and normalization condition of Eqs. (5.1) and (5.2) now take the form

$$\int d\mathbf{x}\, p(\mathbf{x})\, \mathbf{A}(\mathbf{x}) \;=\; \bar{\mathbf{a}} \tag{5.22}$$

$$\int d\mathbf{x}\, p(\mathbf{x}) = 1 \tag{5.23}$$

The PME instructs us to maximize the BGS entropy of Eq. (2.19) subject to the constraints of Eqs. (5.22) and (5.23). This can again be accomplished by means of Lagrange's method of undetermined multipliers, which now requires us to compute the unconstrained maximum of the functional

$$S_\alpha[p(\mathbf{x})] \equiv -\int d\mathbf{x}\, p(\mathbf{x}) \log \frac{p(\mathbf{x})}{\rho(\mathbf{x})} + \alpha_0 \int d\mathbf{x}\, p(\mathbf{x}) + \boldsymbol{\alpha} \cdot \int d\mathbf{x}\, p(\mathbf{x})\, \mathbf{A}(\mathbf{x}) \tag{5.24}$$

with respect to the probability density $p(\mathbf{x})$. To this end, we first compute $\delta S_\alpha \equiv S_\alpha[p(\mathbf{x}) + \delta p(\mathbf{x})] - S_\alpha[p(\mathbf{x})]$ and linearize the result with respect to $\delta p(\mathbf{x})$. We thereby obtain

$$\delta S_\alpha = -\int d\mathbf{x}\, \delta p(\mathbf{x}) \left[\log \frac{p(\mathbf{x})}{\rho(\mathbf{x})} + (1 - \alpha_0) - \boldsymbol{\alpha} \cdot \mathbf{A}(\mathbf{x}) \right] \tag{5.25}$$

The condition for $S_\alpha[p(\mathbf{x})]$ to be a maximum is that $\delta S_\alpha = 0$ for any choice of $\delta p(\mathbf{x})$, which clearly requires

$$\log \frac{p(\mathbf{x})}{\rho(\mathbf{x})} = \alpha_0 - 1 + \boldsymbol{\alpha} \cdot \mathbf{A}(\mathbf{x}) \tag{5.26}$$

The probability density $p(\mathbf{x})$ is therefore given by

$$p(\mathbf{x}) = (1/Z)\, \rho(\mathbf{x}) \exp \boldsymbol{\alpha} \cdot \mathbf{A}(\mathbf{x}) \tag{5.27}$$

where $Z = \exp(1 - \alpha_0)$. Equation (5.27) defines the generalized canonical probability density for continuous states. Equations (5.22), (5.23), and (5.27) combine to imply that

$$Z = Z(\boldsymbol{\alpha}) = \int d\mathbf{x}\, \rho(\mathbf{x}) \exp \boldsymbol{\alpha} \cdot \mathbf{A}(\mathbf{x}) \tag{5.28}$$

and

$$\bar{\mathbf{a}} = (1/Z) \int d\mathbf{x}\, \rho(\mathbf{x})\, \mathbf{A}(\mathbf{x}) \exp \boldsymbol{\alpha} \cdot \mathbf{A}(\mathbf{x}) \tag{5.29}$$

Equations (5.28) and (5.29) are the obvious analogs for continuous states of Eqs. (5.6) and (5.7) for discrete states, and again combine to imply Eq. (5.12) and determine $\bar{\mathbf{a}}$ as a function of $\boldsymbol{\alpha}$, and vice versa. The probability density $p(\mathbf{x})$ thereby depends parametrically on $\bar{\mathbf{a}}$, and will be written as $p(\mathbf{x}|\bar{\mathbf{a}})$ when it is desirable to emphasize this dependence. From this point on the remainder of the formalism for continuous states can simply be obtained from the corresponding relations for discrete states by the use of Eq. (5.21). The density of states and probability density in $\mathbf{A}$-space now become

$$\omega(\mathbf{a}) \equiv \int d\mathbf{x}\, \rho(\mathbf{x})\, \delta(\mathbf{a} - \mathbf{A}(\mathbf{x})) \tag{5.30}$$

$$\begin{aligned} p(\mathbf{a}|\bar{\mathbf{a}}) &\equiv \int d\mathbf{x}\; p(\mathbf{x})\; \delta(\mathbf{a} - \mathbf{A}(\mathbf{x})) \\ &= (1/Z)\, \omega(\mathbf{a}) \exp \boldsymbol{\alpha}\cdot\mathbf{a} \end{aligned} \tag{5.31}$$

which are no longer discontinuous. Equation (5.8) now becomes

$$\langle \mathbf{F}(\mathbf{A}(\mathbf{x})) \rangle \equiv \int d\mathbf{x}\, p(\mathbf{x})\, \mathbf{F}(\mathbf{A}(\mathbf{x})) = \int d\mathbf{a}\, p(\mathbf{a}|\bar{\mathbf{a}})\, \mathbf{F}(\mathbf{a}) \tag{5.32}$$

while Eq. (5.17) is replaced by

$$S(\bar{\mathbf{a}}) = -\int d\mathbf{x}\, p(\mathbf{x}|\bar{\mathbf{a}}) \log \frac{p(\mathbf{x}|\bar{\mathbf{a}})}{\rho(\mathbf{x})} \tag{5.33}$$

Equations (5.10), (5.12), (5.15), (5.16), and (5.18)–(5.20) remain valid as they stand for continuous states.

5.3 A Premonition of Thermodynamics

As previously mentioned, the most important observable in physical systems is the energy, so it is of particular interest to specialize the general development above to the case in which the energy E is presumed to be the only relevant state variable. Thus we set $\mathbf{A} = E$, and we shall regard the microstates as discrete, so that E_k is the energy of the system in microstate k. At the risk of being overly ambitious, we shall presume to consider a macroscopic thermodynamic system of N identical particles (atoms or molecules) confined to a fixed volume V in thermal equilibrium with a heat bath or reservoir at absolute temperature T. The microstate energies E_k will then depend parametrically on N and V, so that dependence will be understood in what follows even when it is not explicitly indicated by the notation. There is no similar dependence on T because the coupling between the system and the heat bath is presumed to be so small that it does not affect the energies of the microstates. As is well known, the parameters (N, V, T) suffice to uniquely determine the thermodynamic macrostate of the system, including its thermodynamic internal energy $\bar{E}$. However, the temperature T is not a microscopic state variable, so it is not yet clear how its value is to be introduced into the formalism. Nevertheless, we shall proceed in the sanguine hope and expectation that this will reveal itself in due course. If the system were isolated its energy would be constant and it would be appropriate to impose the sharp constraint $E_k = E$, or rather $E_k \cong E$ to allow for a small tolerance as discussed in Sects. 3.1 and 4.2.1. However, this would not be appropriate here because the system under consideration is not isolated. In this situation, it seems appropriate to identify the mean value of E_k with the thermodynamic energy $\bar{E}$, and therefore to impose the mean constraint

$$\langle E_k \rangle \;=\; \sum_k p_k E_k \;=\; \bar{E} \tag{5.34}$$

The general development of Sect. 5.2.1 can be specialized to the present situation simply by setting $\mathbf{A}_k = E_k$, $\bar{\mathbf{a}} = \bar{E}$, and $\boldsymbol{\alpha} = \alpha \equiv -\beta$, where the minus sign has been introduced because β will ultimately be found to be positive. Equations (5.5)–(5.7), (5.12), (5.16), and (5.18) then reduce to

$$p_k \;=\; (1/Z)\exp(-\beta E_k) \tag{5.35}$$

$$Z \;=\; Z(\beta) = \sum_k \exp(-\beta E_k) \tag{5.36}$$

$$\bar{E} = \frac{1}{Z}\sum_k E_k \exp(-\beta E_k) = -\frac{\partial \log Z}{\partial \beta} \tag{5.37}$$

$$S = \log Z + \beta \bar{E} \tag{5.38}$$

$$\beta = (\partial S/\partial \bar{E})_{N,V} \tag{5.39}$$

Equation (5.35) is the celebrated Gibbs canonical probability distribution, which is arguably the single most important formula in statistical mechanics. The exponential factor $\exp(-\beta E_k)$ therein is often called the Boltzmann factor.

Equations (5.36) and (5.37) combine to determine $\bar{E}$ as a function of β, or vice versa, with an additional parametric dependence on N and V as noted above. In thermodynamics, however, $\bar{E}$ is a definite function of (N, V, T), which implies the existence of a functional relation between β and T. We further recall that the thermodynamic entropy S_ϑ satisfies the identity $(\partial S_\vartheta/\partial \bar{E})_{N,V} = 1/T$, which bears a suggestive resemblance to Eq. (5.39). Indeed, a comparison between the two shows that if S_ϑ were simply proportional to the statistical entropy S, so that $S_\vartheta = \kappa S$ for some constant κ, then the relation $\beta = 1/(\kappa T)$ would immediately obtain. This observation is highly suggestive, but is not yet sufficient to conclusively establish the identification $S_\vartheta = \kappa S$. This identification will be confirmed in the next chapter by means of more detailed and persuasive arguments based directly on the BP entropy and the EAPP hypothesis rather than the BGS entropy and the PME. The essential equivalence between the two approaches will then become evident in Chapter 7. Conversely, once κS is identified with S_ϑ the PME is further reinforced by the fact that it is the precise statistical analog of the thermodynamic principle that S_ϑ is a maximum at constant $(N, V, \bar{E})$ [29].

It is straightforward to generalize the above development to a system in simultaneous thermal and diffusional equilibrium with a reservoir with which it is allowed to exchange particles as well as energy, so that the number of particles N is no longer a specified constant but now becomes variable. The chemical potential μ of the reservoir then enters the narrative as the parameter which determines the mean number of particles $\bar{N}$ in the same way that T determines the mean energy $\bar{E}$. Thus the thermodynamic state of the system is now defined by the independent thermodynamic variables (μ, V, T), of which $\bar{E}$ and $\bar{N}$ are definite functions. To perform this generalization, it is essential to note that the index k labels the microstates for a given fixed value of N, so a complete description of the microstate of a system in which N can vary requires specification of both N and k. Thus the microstates of the system must now be labeled by the composite index (N, k), which can be effected by means of the replacement $k \to (N, k)$. At the same time, the quantity N is itself a state variable $N_{(N,k)} \equiv N_{Nk}$, whose numerical value in microstate (N, k) is simply $N_{Nk} = N$. In the present context, the mean constraints of Eq. (5.1) then take the form

$$\langle E_{Nk} \rangle = \sum_{Nk} p_{Nk} E_{Nk} = \bar{E} \tag{5.40}$$

$$\langle N \rangle = \sum_{Nk} p_{Nk} N = \bar{N} \tag{5.41}$$

The general development of Sect. 5.2.1 can now be specialized to the present situation by setting $\mathbf{A}_{Nk} = (E_{Nk}, N_{Nk}) = (E_{Nk}, N)$, $\bar{\mathbf{a}} = (\bar{E}, \bar{N})$, and $\boldsymbol{\alpha} = (\alpha_E, \alpha_N) \equiv (-\beta, -\eta)$. Equations (5.5)–(5.7), (5.12), (5.16), and (5.18) then reduce to

$$p_{Nk} = (1/\Xi) \exp(-\beta E_{Nk} - \eta N) \tag{5.42}$$

$$\Xi = \Xi(\beta, \eta) = \sum_{Nk} \exp(-\beta E_{Nk} - \eta N) \tag{5.43}$$

$$\bar{E} = \frac{1}{\Xi} \sum_{Nk} E_{Nk} \exp(-\beta E_{Nk} - \eta N) = -\frac{\partial \log \Xi}{\partial \beta} \tag{5.44}$$

$$\bar{N} = \frac{1}{\Xi} \sum_{Nk} N \exp(-\beta E_{Nk} - \eta N) = -\frac{\partial \log \Xi}{\partial \eta} \tag{5.45}$$

$$S = \log \Xi + \beta \bar{E} + \eta \bar{N} \tag{5.46}$$

$$\beta = (\partial S / \partial \bar{E})_{\bar{N},V} \tag{5.47}$$

$$\eta = (\partial S / \partial \bar{N})_{V,\bar{E}} \tag{5.48}$$

Equation (5.42) is the Gibbs grand canonical probability distribution, and Ξ is the grand canonical partition function. The exponential factor $\exp(-\eta N - \beta E_{Nk})$ is sometimes called the Gibbs factor.

Equations (5.43)–(5.45) combine to determine $\bar{E}$ and $\bar{N}$ as functions of β and η, which can in principle be inverted to express β and η as functions of $\bar{E}$ and $\bar{N}$. (Of course, all those functions also retain an implicit dependence on the fixed parameter V.) In thermodynamics, $\bar{E}$ and $\bar{N}$ can be expressed as functions of (μ, V, T), which implies the existence of a functional relation between η and (μ, T) in addition to the relation between β and T discussed above. We now observe that the thermodynamic identity $(\partial S_\vartheta / \partial \bar{N})_{V,\bar{E}} = -\mu/T$ resembles Eq. (5.48), to which it would become identical if $S_\vartheta = \kappa S$ and $\eta = -\mu/(\kappa T)$. This observation lends further credence to the tentative identification of κS with S_ϑ, although the coefficient of proportionality κ as yet remains undetermined.

Equations (5.42)–(5.46) assume a more transparent and familiar form if we transform from the variable η to the variable $\lambda = \exp(-\eta)$. We thereby obtain

$$p_{Nk} = (1/\Xi)\, \lambda^N \exp(-\beta E_{Nk}) \tag{5.49}$$

$$\Xi = \Xi(\beta, \lambda) = \sum_N \lambda^N Z_N(\beta) \tag{5.50}$$

$$\bar{E} = -\frac{\partial \log \Xi}{\partial \beta} \tag{5.51}$$

$$\bar{N} = \frac{\partial \log \Xi}{\partial \log \lambda} \tag{5.52}$$

$$S = \log \Xi + \beta \bar{E} - \bar{N} \log \lambda \tag{5.53}$$

where $Z_N \equiv \sum_k \exp(-\beta E_{Nk})$ is simply the canonical partition function for a system containing N particles. Equations (5.49)–(5.53) summarize the well known essential formulae of the Gibbs grand canonical ensemble, which have here been obtained as straightforward implications of the PME.

5.4 Jaynes' Proof of the Second Law

We are now in a position to address the apparently paradoxical result obtained at the end of Chapter 3, namely that the BGS entropy S of Eq. (2.19) is constant in time for a wide class of systems with continuous states $\mathbf{x}$ which evolve in time according to deterministic dynamical equations of the form of Eq. (3.3). This class of systems includes the Hamiltonian systems which serve as the usual basis for the statistical mechanics of thermodynamic systems, in which the thermodynamic entropy S_ϑ is expected to increase in accordance with the second law of thermodynamics. Of course, we have not yet established a definite relation between S and S_ϑ, but the PME suggests that in its capacity as a measure of uncertainty, the statistical entropy S would be expected to increase in its own right during the course of a natural time evolution. It is therefore important to understand why this does not at first seem to occur, and how such an increase actually does manifest itself when the formalism is properly interpreted.

For many decades, the conventional approach to resolving this apparent paradox has been to introduce additional uncertainty into the probability distribution by a procedure known as "coarse graining" [17,26], which however is *ad hoc*, imprecise, and lacks a satisfactory theoretical basis [36,37]. An elegant alternative resolution based on the PME was subsequently presented by Jaynes [36]. Jaynes' proof remains less well known than it deserves to be, possibly because its mathematical triviality belies its conceptual subtlety, and the concise form in which Jaynes presented the argument might perhaps have been more transparent. We shall therefore provide a more detailed rendition of the proof.

The critical point which must be clearly appreciated is that the statistical formalism actually provides two different expressions for the entropy as a function of time. Those expressions are not equivalent, and only one of them corresponds to and properly reflects the expected increase in uncertainty. The first such expression is simply the BGS entropy of the probability density $p(\mathbf{x}, t)$ which satisfies the generalized Liouville Eq. (3.7); i.e.,

$$S(t) = -\int d\mathbf{x}\, p(\mathbf{x}, t) \log \frac{p(\mathbf{x}, t)}{\rho(\mathbf{x})} \tag{5.54}$$

As shown in Sect. 3.2.2, $S(t)$ is actually independent of time; i.e., $S(t) = S(0)$ for all t. However, the time evolution of $p(\mathbf{x}, t)$ itself as determined by Eq. (3.7) implies a corresponding time dependence of the mean values of the observables $\mathbf{A}(\mathbf{x})$, which are given by

$$\bar{\mathbf{a}}(t) = \int d\mathbf{x}\, p(\mathbf{x}, t)\, \mathbf{A}(\mathbf{x}) \tag{5.55}$$

and are not in general constant in time unless the variables $\mathbf{A}(\mathbf{x})$ are constants of the motion.

The second expression for the entropy as a function of time is $S(\bar{\mathbf{a}}(t))$, where the function $S(\bar{\mathbf{a}})$ is defined by Eq. (5.33). Thus $S(\bar{\mathbf{a}}(t))$ is simply the BGS entropy of the generalized canonical probability density $p(\mathbf{x}|\bar{\mathbf{a}}(t))$ consistent with the mean constraints

$$\bar{\mathbf{a}}(t) = \int d\mathbf{x}\, p(\mathbf{x}|\bar{\mathbf{a}}(t))\, \mathbf{A}(\mathbf{x}) \tag{5.56}$$

In general $S(t) \neq S(\bar{\mathbf{a}}(t))$, and in contrast to $S(t)$ there are no grounds for expecting $S(\bar{\mathbf{a}}(t))$ to be independent of time. By construction, $p(\mathbf{x}|\bar{\mathbf{a}}(t))$ is the probability density which maximizes the BGS entropy subject to the mean constraints of Eq. (5.56), so $S(\bar{\mathbf{a}}(t))$ is the largest entropy consistent with those constraints. But Eq. (5.55) shows that $p(\mathbf{x}, t)$ satisfies precisely the same constraints, which implies that $S(t)$ cannot exceed $S(\bar{\mathbf{a}}(t))$. Thus

$$S(\bar{\mathbf{a}}(t)) \geq S(t) = S(0) \tag{5.57}$$

According to Eq. (5.54),

$$S(0) = -\int d\mathbf{x}\, p(\mathbf{x}, 0) \log \frac{p(\mathbf{x}, 0)}{\rho(\mathbf{x})} \tag{5.58}$$

and in the absence of other information the initial probability density $p(\mathbf{x}, 0)$ is normally defined to be the generalized canonical probability density consistent with the initial values $\bar{\mathbf{a}}(0)$. Thus we set $p(\mathbf{x}, 0) = p(\mathbf{x}|\bar{\mathbf{a}}(0))$ in Eq. (5.58), which then combines with Eq. (5.33) to imply $S(0) = S(\bar{\mathbf{a}}(0))$. Equation (5.57) then becomes

$$S(\bar{\mathbf{a}}(t)) \geq S(\bar{\mathbf{a}}(0)) \tag{5.59}$$

If the system approaches statistical equilibrium as $t \to \infty$ then the variables $\bar{\mathbf{a}}(t)$ approach well defined limiting values $\bar{\mathbf{a}}(\infty)$, and Eq. (5.59) then implies

$$S(\bar{\mathbf{a}}(\infty)) \geq S(\bar{\mathbf{a}}(0)) \tag{5.60}$$

Equation (5.60) shows that the final entropy $S(\bar{\mathbf{a}}(\infty))$ cannot be smaller, and will ordinarily be larger, than the initial entropy $S(\bar{\mathbf{a}}(0))$, just as one would intuitively expect. However, Eq. (5.59) does not imply that the increase from $S(\bar{\mathbf{a}}(0))$ to $S(\bar{\mathbf{a}}(\infty))$ occurs monotonically. Note that this derivation makes no reference to the thermodynamic entropy S_ϑ, and indeed is not restricted to macroscopic thermodynamic systems. Equations (5.59) and (5.60) are therefore valid in general, regardless of whether or not S is related to S_ϑ. Once the identification of κS with S_ϑ has been established, however, Eq. (5.60) confirms that the statistical interpretation of entropy is indeed consistent with the second law of thermodynamics.

Although Eq. (5.59) is mathematically rigorous, it may not be intuitively obvious that $S(\bar{\mathbf{a}}(t))$ rather than $S(t)$ is the more appropriate entropy to evaluate and scrutinize in time-dependent problems. The rationale for this premise becomes clearer when one contemplates the uncertainty which each of those entropies represents as to the state of the system at time t. The initial condition $p(\mathbf{x}, 0) = p(\mathbf{x}|\bar{\mathbf{a}}(0))$ implies that $S(0) = S(t)$ represents the maximum uncertainty consistent with the *initial* mean values $\bar{\mathbf{a}}(0)$, whereas $S(\bar{\mathbf{a}}(t))$ represents the maximum uncertainty consistent with the *current* mean values $\bar{\mathbf{a}}(t)$ at time t. Those two uncertainties differ because maximum uncertainty is not in general preserved by Eq. (3.7); if $p(\mathbf{x}, 0)$ has the generalized canonical form of Eq. (5.27) then $p(\mathbf{x}, t)$ generally does not, and vice versa. Thus $S(\bar{\mathbf{a}}(t))$ represents the maximum uncertainty as to the state of the system at time t consistent with the values $\bar{\mathbf{a}}(t)$ alone, regardless of the form of $p(\mathbf{x}, 0)$. (Of course, $p(\mathbf{x}, 0)$ uniquely determines $\bar{\mathbf{a}}(t)$ via Eqs. (3.7) and (5.55), but $\bar{\mathbf{a}}(t)$ does not conversely determine $p(\mathbf{x}, 0)$ uniquely.) As such, $S(\bar{\mathbf{a}}(t))$ is inherently local in time with no history dependence, and thereby provides an appropriate intrinsic measure of the instantaneous uncertainty as to the state of the system at time t. We also note that defining the time-dependent entropy as $S(\bar{\mathbf{a}}(t))$ is entirely analogous to the local equilibrium hypothesis in nonequilibrium thermodynamics, whereby the entropy in orthodox classical thermodynamics (which is defined only in equilibrium and corresponds to the function $S(\bar{\mathbf{a}})$ in the present context) is phenomenologically extended to nonequilibrium processes in which the thermodynamic state variables upon which it depends vary slowly with time.

Chapter 6

THERMODYNAMIC SYSTEMS

In the preceding chapters, the mathematical fundamentals of entropy have been deliberately developed in a general form in which the nature of the system and its states have been left unspecified, with only occasional intimations (mainly in Sect. 5.3) of the possible physical applications of the formalism. This approach was adopted in order to preserve maximum generality for as long as possible, thereby making it easier to clearly distinguish between general features and those which are specific or peculiar to a particular context. For example, the treatment of constraints in Chapter 4 shows that Einstein fluctuation formulae, which are commonly used to describe thermodynamic fluctuations, are not restricted to thermodynamic systems but are actually more general. We now proceed to systematically explore the application of the preceding general concepts to macroscopic thermodynamic systems. The brief prelude to such systems in Sect. 5.3 was merely an hors d'oeuvre based on the BGS entropy and the PME, which was made plausible but in a logical sense constitutes an additional postulate. In contrast, the more detailed treatment in this chapter is based entirely on the BP entropy and the EAPP hypothesis, and is thereby more fundamental with fewer assumptions. As we shall see in Chapter 7, however, both approaches ultimately lead to essentially the same results, a circumstance which provides compelling further evidence in support of the PME.

Macroscopic thermodynamic systems contain very large numbers ($\sim 10^{23}$) of very small particles (atoms or molecules). Each possible internal arrangement or configuration of those particles constitutes a microscopic state or *microstate* of the macroscopic many-particle system. As discussed in most textbooks (e.g., [38]), the number of microstates generally increases exponentially with the number of particles N and is consequently much larger yet than N itself. Indeed, the number of microstates of a macroscopic system is so incomprehensibly enormous that it lies entirely outside the domain of our ordinary everyday experience and intuition. Thus it is well to prepare ourselves in advance for possibly counterintuitive surprises. It is by no means trivial or obvious how to properly define, characterize, and enumerate such microstates. This task will be undertaken in Chapter 8. The scope of the present chapter is limited to general relations which depend in essence only on the existence of a vast multitude of microstates, and which can therefore be formally established without considering their detailed structure or characteristics. It is most remarkable that even in the absence of those details, the resulting general relations are nevertheless sufficient to strongly suggest and lend credence to the proposition that the statistical and thermodynamic entropies are essentially one and the same.

The qualitative nature of the microstates of a many-particle system depends on the physical theory used to describe the particles. For finite systems, the microstates are continuous in classical mechanics but discrete in quantum mechanics. Quantum mechanics is the more fundamental theory, and discrete states are in many ways simpler to deal with than continuous states, so the development in this chapter presumes that the microstates are discrete. The classical description of multi-particle systems in terms of continuous microstates will be discussed in Chapter 8.

As discussed in Chapter 4, in most situations of physical interest the number of microstates W and their probabilities p_k are not fixed constants but rather depend on the values of a relatively small number of externally controlled parameters, which normally represent and correspond to constraints imposed on the system. In macroscopic thermodynamic systems, those constraint parameters are the independent thermodynamic state variables that define the macrostate of the system, as described in Sect. 2.2. The full machinery of constraints as developed in Chapter 4 is not required for present purposes, but much of the following development essentially constitutes a particular special case of that more general discussion.

6.1 Isolated Systems

We shall initially restrict attention to homogeneous systems of N identical particles confined to a volume V with total energy E. It is well established experimentally that such systems possess equations of state which reproducibly express other thermodynamic variables (e.g., the pressure P and temperature T) as definite functions of (N, V, E). The macroscopic variables (N, V, E) therefore suffice to uniquely determine the thermodynamic state of the system, and they thereby define the *macrostate* of the system in the present context.

The thermodynamic variables of sufficiently large systems can usually be classified as extensive or intensive according as their functional dependence on (N, V, E) is of the form $Nf(V/N, E/N)$ or $f(V/N, E/N)$, respectively. Thus N, V, and E themselves are all extensive, while $P, T, V/N$, and E/N are intensive. When the above functional dependences obtain, the thermodynamic equations of state can be expressed entirely in terms of intensive variables, which then reduces the number of independent variables for pure systems from three to two. However, the functional forms given above are not sufficiently general to accommodate all systems and situations of interest (e.g., surface effects, long-range forces), and in any case a complete description of the state of the system must include at least one extensive variable to determine the size of the system, which would otherwise be indeterminate. Moreover, the concept of extensivity plays no role in the general theory, and when it does occur it automatically emerges from the formalism and need not be assumed.

Thus we wholly concur with those who do not consider extensivity an essential feature of either entropy or thermodynamics, and we retain the full general functional dependence on the three independent variables (N, V, E). We employ those particular variables rather than some other thermodynamically equivalent set such as (N, V, T) because (a) N, V, and E are all mechanical variables which are well defined independently of thermodynamics, and (b) it is conceptually easiest to envision and deal with isolated systems, for which the energy E is a conserved variable with a constant value that does not fluctuate. Statistical quantities that correspond to various other common thermodynamical variables such as P and T will emerge naturally as we proceed.

The preceding discussion suggests that the number of microstates $W(N, V, E)$ consistent with a given macrostate (N, V, E) is expected to be enormous. According to quantum mechanics, however, each discrete microstate k has a definite energy $E_k(N, V)$ (which depends parametrically on N and V as the notation indicates). Even if those discrete energy values are extremely numerous and closely spaced, as indeed they are for macroscopic systems, they nevertheless remain discrete, whereas the thermodynamic energy E of the macroscopic system is a continuous variable. Thus the number of microstates that are *precisely* consistent with any particular given values of (N, V, E) is almost always zero, because almost all values of the continuous variable E differ from *any* of the discrete values $E_k(N, V)$. As discussed in Sects. 3.1 and 4.2.1, this circumstance requires us to introduce a macroscopically small tolerance or uncertainty $\Delta E \ll E$ in the value of E. This can be justified, or at least rationalized, by observing that even though a hypothetical isolated system has a definite sharp value of E in principle, that value cannot be either controlled or measured with infinite precision. For macroscopic purposes ΔE is negligible and can usually be ignored, but in the present context it must be taken into account in defining the number of microstates W and their probabilities p_k as functions of (N, V, E). Thus we define $W(N, V, E)$ as the number of microstates k for which

$|E_k(N,V) - E| < \frac{1}{2}\Delta E$, or

$$E - \tfrac{1}{2}\Delta E < E_k(N,V) < E + \tfrac{1}{2}\Delta E \qquad (6.1)$$

where $\Delta E \ll E$. Equation (6.1) can be more compactly and eloquently expressed as $E_k(N,V) \in \mathbb{N}(E)$, where $\mathbb{N}(E)$ is the neighborhood of E formally defined by

$$\mathbb{N}(E) \equiv \{E' | E - \tfrac{1}{2}\Delta E < E' < E + \tfrac{1}{2}\Delta E\}$$

In the present context, $\mathbb{N}(E)$ corresponds to the neighborhood $\mathbb{N}(\mathbf{a})$ in the general treatment of Chapter 4, and the tolerance ΔE corresponds to its volume $v(\mathbf{a})$. Since E is an additive variable, ΔE is taken to be a constant independent of E, as discussed in Sect. 4.2.3. The accessible microstates k for given values of N, V, and E are those for which Eq. (6.1) is satisfied. The total number $W(N,V,E)$ of such microstates, and their probabilities $p_k(N,V,E)$, are then well defined in principle, but of course they now depend implicitly on ΔE. In due course, however, we shall see that the resulting entropy is normally insensitive to the choice of ΔE within very wide limits.

Although systems with continuous microstates $\mathbf{x}$ will not be considered in this chapter, it should be noted that a small but finite energy tolerance ΔE is equally essential in that context as well, because the condition $E(\mathbf{x}) = E$ defines a set of measure zero in $\mathbf{x}$-space as discussed in Sect. 4.2.1. In such systems Eq. (6.1) is accordingly replaced by $|E(\mathbf{x}) - E| < \frac{1}{2}\Delta E$ or $E(\mathbf{x}) \in \mathbb{N}(E)$, which defines a finite region $\mathbb{E}$ in $\mathbf{x}$-space and thereby a finite number of microstates $W = \int_{\mathbb{E}} d\mathbf{x}\, \rho(\mathbf{x})$.

Having emphasized that ΔE cannot be too large, we must also emphasize that it cannot be too small either; it must remain large enough for $W(N,V,E)$ to be a very large number. The essential reason for this requirement is that $W(N,V,E)$ is in principle an integer and is consequently a discontinuous function of E, but in the present context it must be treated as a continuous function for practical purposes. This requires that its discontinuities be small compared to its value, which in turn requires that ΔE be large enough to encompass a very large number of microstates. Thus ΔE must greatly exceed the typical spacing between adjacent microstate energies E_k, which is naturally expected to be of a microscopic or molecular order of magnitude associated with the individual particles. In short, ΔE must be small in a macroscopic sense but large in a microscopic or molecular sense.

The number of microstates $W(N, V, E)$ consistent with Eq. (6.1) can be formally expressed as

$$W(N, V, E) = \int_{\mathbb{N}(E)} dE' \, \omega(E'|N, V) \tag{6.2}$$

where

$$\omega(E|N, V) \equiv \sum_k \delta(E - E_k(N, V))$$

is the true discontinuous density of states of the system in energy space, which has the property that its integral over any energy interval is the number of microstates k with energies therein. The mean or average density of states within the energy interval ΔE is simply

$$\bar{\omega}(E|N, V) \equiv (1/\Delta E) \, W(N, V, E)$$

so that

$$W(N, V, E) = \bar{\omega}(E|N, V) \, \Delta E \tag{6.3}$$

Thus $\bar{\omega}(E|N, V)$, like $W(N, V, E)$, is discontinuous in principle but can be treated as continuous in the same way and to the same degree of approximation. In addition to being very numerous and closely spaced, the microstate energies $E_k(N, V)$ are presumed to be so smoothly and densely distributed in E that $W(N, V, E)$ and hence $\bar{\omega}(E|N, V)$ are very nearly constant over energy intervals of order ΔE. It then follows that

$$W(N, V, E) = \bar{\omega}(E|N, V) \, \Delta E \cong \int_{\mathbb{N}(E)} dE' \, \bar{\omega}(E'|N, V)$$

so that each differential energy element dE can be regarded as containing $\bar{\omega}(E|N,V)\,dE$ microstates. Comparison with Eq. (6.2) then shows that $\bar{\omega}(E|N,V)$ constitutes a smoothed continuous approximation to $\omega(E|N,V)$, in the sense that the number of microstates within any energy interval $\gtrsim \Delta E$ can be accurately approximated by integrating $\bar{\omega}(E|N,V)$ over that interval. This situation is frequently described by saying that the microstate energies $E_k(N,V)$ form a quasi-continuous spectrum. It is obviously much more convenient to deal with the continuous mean density of states $\bar{\omega}(E|N,V)$ than the true discontinuous density of states $\omega(E|N,V)$, so it will henceforth be understood that the term "density of states" refers to the former rather than the latter. In what follows, the parametric arguments N and/or V will often be suppressed for simplicity in situations where they do not vary. Simple estimates based on ideal systems show that $\bar{\omega}(E)$ typically increases monotonically and extremely rapidly with E. For example, $\bar{\omega}(E) \sim E^{N-1}$ and $E^{3N/2-1}$ for harmonic oscillators and monatomic ideal gases, respectively [38].

We now invoke the EAPP hypothesis by assuming that all of the $W(N,V,E)$ accessible microstates k defined by Eq. (6.1) are equally probable, so that $p_k(N,V,E) = 1/W(N,V,E)$ independently of k. We reemphasize that the EAPP hypothesis in the present context is inherently specific to and contingent upon the fact that the system is isolated so that its energy is constant, as discussed in Chapter 3. The entropy of the system is then simply the BP entropy of Eq. (2.8),

$$S(N,V,E) = \log W(N,V,E) = \log(\bar{\omega}(E)\Delta E) \qquad (6.4)$$

Of course, we have not as yet established any relation between S and the thermodynamic entropy of the system, so at this point S simply represents the uncertainty as to which microstate the system occupies. Since $W(N,V,E)$ and $\bar{\omega}(E)$ are regarded as continuous functions of E, the same is true of $S(N,V,E)$.

The aforementioned insensitivity of S to ΔE can now be understood in terms of the logarithmic accuracy considerations discussed at the end of Sect. 2.3. For this purpose it is useful to introduce the total number of microstates with energies $E_k \leq E$, which is given by

$$\Omega(E) = \int_0^E dE' \,\varpi(E') \tag{6.5}$$

Thus

$$\varpi(E) = \partial\Omega/\partial E$$

The argument is based on the obvious inequalities

$$\varpi(E)\Delta E \leq \Omega(E) \leq E\,\varpi(E) \tag{6.6}$$

which combine with Eq. (6.4) to imply

$$S \leq \log\Omega(E) \leq S + \log(E/\Delta E) \tag{6.7}$$

For any reasonable macroscopic tolerance ΔE, even very small ones such as $\Delta E = 10^{-10}E$, $\log(E/\Delta E)$ will exceed unity by at most a few orders of magnitude. In contrast, $S = \log(\varpi(E)\Delta E)$ is typically of order $N \sim 10^{23}$, because the number of microstates typically increases exponentially with N as previously discussed. It follows that $\log(E/\Delta E)$ is utterly negligible compared to S, whereupon Eq. (6.7) reduces to the equality

$$S = \log\Omega(E) \tag{6.8}$$

which is a more convenient expression for S than Eq. (6.4), and explicitly manifests its independence of ΔE. Thus, even though S depends on ΔE in principle, it exhibits a plateau value which is very nearly constant over the wide range of values for which ΔE is physically sensible. The fact that the superficially very different Eqs. (6.4) and (6.8) are essentially equivalent shows that only microstates in the immediate vicinity of E contribute significantly to $\log\Omega(E)$ and thence to S. Note, however, that this does *not* imply that almost all microstates with $E_k \leq E$ have energies in the immediate vicinity of E. Once again the relevant concept is logarithmic accuracy; e.g., 10^{303} is 1000 times larger than 10^{300}, but their logarithms differ by only 1%. The microstates with $E_k \cong E$ do not necessarily dominate those with $E_k \leq E$ in an additive sense, but only in a logarithmic or multiplicative sense, which however is all that is necessary for purposes of computing S.

Most textbooks further state that the equation

$$S = \log \bar{\omega}(E)$$

is also essentially equivalent to Eqs. (6.4) and (6.8), but that equation is nonsensical as written because $\bar{\omega}(E)$ is not dimensionless. However, its dimensionless numerical value in any particular system of units is simply $\widetilde{\omega}(E) = \hat{e}\,\bar{\omega}(E)$, where $\hat{e}$ is the unit of energy, so Eq. (6.4) can be rewritten as

$$S = \log(\widetilde{\omega}(E)\Delta E/\hat{e}) = \log \widetilde{\omega}(E) - \log(\hat{e}/\Delta E) \qquad (6.9)$$

It should now be clear from the preceding discussion that the term $\log(\hat{e}/\Delta E)$ is utterly negligible compared to S for any reasonable system of units. Equation (6.9) therefore reduces to

$$S = \log \widetilde{\omega}(E) \qquad (6.10)$$

which now really is essentially equivalent to Eqs. (6.4) and (6.8), and has the paradoxical property of being equally valid in any reasonable system of units. A change in units merely multiplies $\widetilde{\omega}(E)$ by a dimensionless constant factor whose logarithm is completely negligible in comparison to S.

6.2 Systems in Thermal Equilibrium

Now consider two isolated statistically independent thermodynamic systems $\mathcal{A}$ and $\mathcal{B}$, which may be composed of either the same or different types or species of particles, and which respectively occupy the macrostates $(N_{\mathcal{A}}, V_{\mathcal{A}}, E_{\mathcal{A}})$ and $(N_{\mathcal{B}}, V_{\mathcal{B}}, E_{\mathcal{B}})$. The microstates of systems $\mathcal{A}$ and $\mathcal{B}$ are respectively labeled by the discrete indices i and j, and their corresponding energies are denoted by $E_i^{\mathcal{A}}$ and $E_j^{\mathcal{B}}$. According to Eq. (6.3), the numbers of those microstates that lie in the intervals

$$E_{\mathcal{A}} - \tfrac{1}{2}\Delta E \;<\; E_i^{\mathcal{A}} < E_{\mathcal{A}} + \tfrac{1}{2}\Delta E \tag{6.11}$$

$$E_{\mathcal{B}} - \tfrac{1}{2}\Delta E \;<\; E_j^{\mathcal{B}} < E_{\mathcal{B}} + \tfrac{1}{2}\Delta E \tag{6.12}$$

are given by

$$W_{\mathcal{A}} = W_{\mathcal{A}}(N_{\mathcal{A}}, V_{\mathcal{A}}, E_{\mathcal{A}}) = \varpi_{\mathcal{A}}(E_{\mathcal{A}})\Delta E \tag{6.13}$$

$$W_{\mathcal{B}} = W_{\mathcal{B}}(N_{\mathcal{B}}, V_{\mathcal{B}}, E_{\mathcal{B}}) = \varpi_{\mathcal{B}}(E_{\mathcal{B}})\Delta E \tag{6.14}$$

so the entropies of systems $\mathcal{A}$ and $\mathcal{B}$ are given by

$$S_{\mathcal{A}}(E_{\mathcal{A}}) = \log W_{\mathcal{A}}(E_{\mathcal{A}}) = \log(\varpi_{\mathcal{A}}(E_{\mathcal{A}})\Delta E) \tag{6.15}$$

$$S_{\mathcal{B}}(E_{\mathcal{B}}) = \log W_{\mathcal{B}}(E_{\mathcal{B}}) = \log(\varpi_{\mathcal{B}}(E_{\mathcal{B}})\Delta E) \tag{6.16}$$

The use of the same energy tolerance ΔE for both systems is a convenient simplification, and also ensures that systems $\mathcal{A}$ and $\mathcal{B}$ are treated on an equal footing, in the sense that their microstates have equal weight and can be consistently counted and compared. Conversely, the use of different energy tolerances for the two systems would introduce an arbitrary artificial asymmetry between them, although this would not significantly affect their entropies, which as discussed above are insensitive to the choice of ΔE.

Just as was done in Sect. 2.3, systems $\mathcal{A}$ and $\mathcal{B}$ can be regarded as two separated parts of a single combined system $\mathcal{AB}$. The total energy of system $\mathcal{AB}$ is $E = E_{\mathcal{A}} + E_{\mathcal{B}}$, its microstates are indexed by the ordered pairs $k = (i, j)$, and their energies $E_k^{\mathcal{AB}} = E_{ij}^{\mathcal{AB}} = E_i^{\mathcal{A}} + E_j^{\mathcal{B}}$ are constrained by the sum of Eqs. (6.11) and (6.12) to lie in the range

$$E - \Delta E < E_{ij}^{\mathcal{AB}} < E + \Delta E \tag{6.17}$$

The total number of microstates of the combined system for given values of $E_{\mathcal{A}}$ and $E_{\mathcal{B}}$ is simply

$$W_{\mathcal{AB}}(E_{\mathcal{A}}, E_{\mathcal{B}}) \equiv W_{\mathcal{A}} W_{\mathcal{B}} = \bar{\omega}_{\mathcal{A}}(E_{\mathcal{A}}) \bar{\omega}_{\mathcal{B}}(E_{\mathcal{B}}) \Delta E^2 \qquad (6.18)$$

The probabilities of the individual microstates are given by $p_i^{\mathcal{A}} = 1/W_{\mathcal{A}}$, $p_j^{\mathcal{B}} = 1/W_{\mathcal{B}}$, and $p_{ij}^{\mathcal{AB}} = p_i^{\mathcal{A}} p_j^{\mathcal{B}} = 1/W_{\mathcal{AB}}$, so the microstates (i, j) of the combined system $\mathcal{AB}$ are also equally probable. The entropy of the combined system $\mathcal{AB}$ is therefore simply the BP entropy

$$S_0(E_{\mathcal{A}}, E_{\mathcal{B}}) = \log W_{\mathcal{AB}}(E_{\mathcal{A}}, E_{\mathcal{B}}) = S_{\mathcal{A}}(E_{\mathcal{A}}) + S_{\mathcal{B}}(E_{\mathcal{B}}) \qquad (6.19)$$

where subscript zero denotes and reminds us that up to this point systems $\mathcal{A}$ and $\mathcal{B}$ remain separated and noninteracting. Equation (6.19) shows that their separate entropies are indeed additive, just as they should be.

We now imagine that systems $\mathcal{A}$ and $\mathcal{B}$ are placed in thermal contact so that they can exchange energy, but not particles or volume. Both observations and thermodynamics inform us that the exchange of energy will cause the two systems to come into thermal equilibrium with each other. We are not concerned with the dynamical process by which this equilibration occurs, but only with the statistical description of the resulting equilibrium situation in terms of the microstates of the individual systems. The coupling between the two systems is presumed to be sufficiently weak that their interaction energy is negligible, so that the relation $E = E_{\mathcal{A}} + E_{\mathcal{B}}$ remains valid. The exchange of energy does not affect E, which remains fixed at its initial value, but it now allows $E_{\mathcal{A}}$ and $E_{\mathcal{B}} = E - E_{\mathcal{A}}$ to vary between zero and E. Thus the constraints of Eqs. (6.11) and (6.12) no longer apply, so the accessible microstates of systems $\mathcal{A}$ and $\mathcal{B}$ can no longer be presumed to be equally probable. In contrast, the accessible microstates (i, j) of the combined system remain constrained by Eq. (6.17), and therefore remain equally probable according to the EAPP hypothesis. However, those microstates are now more numerous than they were before, because microstates that violate Eqs. (6.11) and (6.12) but satisfy Eq. (6.17) were previously inaccessible but have now become accessible. This implies that the entropy of the system increases as a result of the equilibration, as will be demonstrated in what follows.

Thus the immediate essential effect of placing systems $\mathcal{A}$ and $\mathcal{B}$ in thermal contact is to replace the two constraints of Eqs. (6.11) and (6.12) by the single constraint of Eq. (6.17), which allows $E_{\mathcal{A}}$ and $E_{\mathcal{B}}$ to vary as described above. For a given fixed value of the total energy E, the value of $E_{\mathcal{B}} = E - E_{\mathcal{A}}$ is determined by $E_{\mathcal{A}}$, so it will henceforth be convenient to regard the various quantities defined above as functions of $(E_{\mathcal{A}}, E)$ rather than $(E_{\mathcal{A}}, E_{\mathcal{B}})$. Since E does not vary during the energy exchange and equilibration, the arguments $(E_{\mathcal{A}}, E)$ will be written as $(E_{\mathcal{A}}|E)$ to emphasize that they denote a functional dependence on $E_{\mathcal{A}}$ for a given fixed value of E, which now essentially plays the role of a parameter. Thus Eq. (6.18) for the total number of microstates of the combined system for a given value of $E_{\mathcal{A}}$ within a tolerance ΔE can be rewritten as

$$\begin{aligned} W_{\mathcal{AB}}(E_{\mathcal{A}}|E) &= W_{\mathcal{A}}(E_{\mathcal{A}})\, W_{\mathcal{B}}(E - E_{\mathcal{A}}) \\ &= \bar{\omega}_{\mathcal{A}}(E_{\mathcal{A}})\, \bar{\omega}_{\mathcal{B}}(E - E_{\mathcal{A}})\, \Delta E^2 \end{aligned} \tag{6.20}$$

while Eq. (6.19) becomes

$$S_0(E_{\mathcal{A}}|E) = \log W_{\mathcal{AB}}(E_{\mathcal{A}}|E) = S_{\mathcal{A}}(E_{\mathcal{A}}) + S_{\mathcal{B}}(E - E_{\mathcal{A}}) \tag{6.21}$$

For future reference, we note that $S_0(E_{\mathcal{A}}|E)$ can also be interpreted as the conditional entropy of the combined system $\mathcal{AB}$ subject to the constraint of Eq. (6.11); i.e., for a given value of $E_{\mathcal{A}}$.

The mean density of microstates of the combined system $\mathcal{AB}$ in $E_{\mathcal{A}}$ is simply the number of such microstates per unit increment in $E_{\mathcal{A}}$, which is given by

$$\bar{\omega}_{\mathcal{AB}}(E_{\mathcal{A}}|E) \equiv (1/\Delta E)\, W_{\mathcal{AB}}(E_{\mathcal{A}}|E)$$

When $E_{\mathcal{A}}$ is allowed to vary freely, the total number of accessible microstates $W(E)$ for a given fixed value of E is therefore given by

$$W(E) = \int_0^E dE_{\mathcal{A}}\, \bar{\omega}_{\mathcal{AB}}(E_{\mathcal{A}}|E) = \frac{1}{\Delta E} \int_0^E dE_{\mathcal{A}}\, W_{\mathcal{AB}}(E_{\mathcal{A}}|E) \tag{6.22}$$

Since the microstates of the combined system $\mathcal{AB}$ are equally probable, its entropy is simply given by the BP entropy

$$S(E) = \log W(E) = \log \left(\frac{1}{\Delta E} \int_0^E dE_{\mathcal{A}} \, W_{\mathcal{AB}}(E_{\mathcal{A}}|E) \right) \qquad (6.23)$$

We note parenthetically that Eq. (6.22) would be ambiguous if different energy tolerances were used in Eqs. (6.13) and (6.14), because the factor of ΔE would then differ depending on whether the integration was performed over $E_{\mathcal{A}}$ or $E_{\mathcal{B}}$. This confirms the necessity of using a single ΔE for consistency, but in any case such an ambiguity would be logarithmically negligible and thus would not significantly affect $S(E)$.

Since the value of $E_{\mathcal{A}}$ is no longer fixed, it is of interest to consider its mean or smoothed probability density $\bar{p}_{\mathcal{A}}(E_{\mathcal{A}}|E)$, which is defined in such a way that $\bar{p}_{\mathcal{A}}(E_{\mathcal{A}}|E) \, \Delta E$ is the probability that the energy of system $\mathcal{A}$ lies between $E_{\mathcal{A}} - \frac{1}{2}\Delta E$ and $E_{\mathcal{A}} + \frac{1}{2}\Delta E$ in the final equilibrated system $\mathcal{AB}$ with a given fixed value of E. Clearly $\bar{p}_{\mathcal{A}}(E_{\mathcal{A}}|E)$ is proportional to the number of equally probable microstates of system $\mathcal{AB}$ consistent with the value $E_{\mathcal{A}}$, namely $W_{\mathcal{AB}}(E_{\mathcal{A}}|E)$, and of course it must be normalized so that $\int_0^E dE_{\mathcal{A}} \, \bar{p}_{\mathcal{A}}(E_{\mathcal{A}}|E) = 1$. Thus

$$\bar{p}_{\mathcal{A}}(E_{\mathcal{A}}|E) = \frac{W_{\mathcal{AB}}(E_{\mathcal{A}}|E)}{\Delta E \, W(E)} = \frac{\exp S_0(E_{\mathcal{A}}|E)}{\Delta E \, \exp S(E)} \qquad (6.24)$$

where use has been made of Eqs. (6.21)–(6.23). Note that $\bar{p}_{\mathcal{A}}(E_{\mathcal{A}}|E)$ is entirely independent of ΔE, because the factors of ΔE in its numerator and denominator cancel out. The second equality in Eq. (6.24) is an example of an Einstein fluctuation formula of the type discussed in Chapter 4.

Let $E_{\mathcal{A}}^*$ denote the most probable value of $E_{\mathcal{A}}$; i.e., the value of $E_{\mathcal{A}}$ for which $\bar{p}_{\mathcal{A}}(E_{\mathcal{A}}|E)$ is a maximum. Equation (6.24) shows that $W_{\mathcal{AB}}(E_{\mathcal{A}}|E)$ and $S_0(E_{\mathcal{A}}|E)$ likewise attain their maximum values at

$E_{\mathcal{A}} = E_{\mathcal{A}}^*$. Equation (6.22) then shows that $W(E)$ satisfies the obvious inequalities

$$W_{\mathcal{AB}}(E_{\mathcal{A}}^*|E) \leq W(E) \leq (E/\Delta E)\, W_{\mathcal{AB}}(E_{\mathcal{A}}^*|E) \qquad (6.25)$$

which combine with Eqs. (6.21) and (6.23) to imply

$$S_0(E_{\mathcal{A}}^*|E) \leq S(E) \leq S_0(E_{\mathcal{A}}^*|E) + \log(E/\Delta E) \qquad (6.26)$$

As previously discussed, the term $\log(E/\Delta E)$ is again utterly negliglble, so Eq. (6.26) reduces to the equality

$$S(E) = S_0(E_{\mathcal{A}}^*|E) = S_{\mathcal{A}}(E_{\mathcal{A}}^*) + S_{\mathcal{B}}(E - E_{\mathcal{A}}^*) \qquad (6.27)$$

where use has been made of Eq. (6.21). Equation (6.27) shows that (a) the entropy remains additive for macroscopic systems $\mathcal{A}$ and $\mathcal{B}$ in thermal equilibrium, as would be expected; (b) the equilibrium entropies of systems $\mathcal{A}$ and $\mathcal{B}$ have the same values as if the two systems were isolated with the final equilibrated energies $E_{\mathcal{A}} = E_{\mathcal{A}}^*$ and $E_{\mathcal{B}} = E - E_{\mathcal{A}}^*$; and (c) the final unconstrained entropy $S(E)$ of the combined system $\mathcal{AB}$ in thermal equilibrium is equal to the maximum possible value $S_0(E_{\mathcal{A}}^*|E)$ of the constrained entropy $S_0(E_{\mathcal{A}}|E)$. It is natural to interpret this behavior as a spontaneous increase in the entropy from its initial constrained value $S_0(E_{\mathcal{A}}|E)$ to its final maximum value $S(E)$. This illustrates the general principle discussed in Chapter 4 that the entropy never decreases and normally increases when constraints are removed, which is one of the cornerstones of the PME as discussed in Chapter 5.

We know from everyday experience that macroscopic systems, regardless of whether or not they are isolated, do not normally exhibit observable fluctuations in their thermodynamic state variables. Indeed, if they did those variables could not satisfy reproducible deterministic equations of state, and the science of thermodynamics would not exist. This observation implies that fluctuations in $E_{\mathcal{A}}$ and $E_{\mathcal{B}}$ in the unconstrained combined system $\mathcal{AB}$ should be negligible, and if so then $\bar{p}_{\mathcal{A}}(E_{\mathcal{A}}|E)$ must be very sharply peaked around the value $E_{\mathcal{A}} = E_{\mathcal{A}}^*$. This is not immediately obvious from the preceding development, but is a consequence of the fact that according to Eq. (6.20), $W_{\mathcal{AB}}(E_{\mathcal{A}}|E)$ is the product of a rapidly increasing function of $E_{\mathcal{A}}$ and a rapidly decreasing function of $E_{\mathcal{A}}$. The resulting sharply peaked functional dependence on $E_{\mathcal{A}}$ can readily be understood by considering a simple example in which the densities of states of both systems $\mathcal{A}$ and $\mathcal{B}$ have the previously mentioned prototypical ideal form $\varpi(E) \sim E^N$ [38]. Equations (6.20) and (6.24) then combine to imply that

$$\bar{p}_{\mathcal{A}}(E_{\mathcal{A}}|E) = C\, E_{\mathcal{A}}^{N_{\mathcal{A}}} (E - E_{\mathcal{A}})^{N_{\mathcal{B}}}$$

where the normalization factor C is independent of $E_{\mathcal{A}}$. The value $E_{\mathcal{A}}^*$ which maximizes $\bar{p}_{\mathcal{A}}(E_{\mathcal{A}}|E)$ is determined by setting

$$\frac{\partial \bar{p}_{\mathcal{A}}(E_{\mathcal{A}}|E)}{\partial E_{\mathcal{A}}} = 0$$

with the result $E_{\mathcal{A}}^* = N_{\mathcal{A}} E/N$, where $N = N_{\mathcal{A}} + N_{\mathcal{B}}$. The width of $\bar{p}_{\mathcal{A}}(E_{\mathcal{A}}|E)$ in the vicinity of its maximum is of order $\delta E_{\mathcal{A}}^*$, where

$$\frac{1}{(\delta E_{\mathcal{A}}^*)^2} = \left| \frac{\partial^2 \log \bar{p}_{\mathcal{A}}(E_{\mathcal{A}}|E)}{\partial E_{\mathcal{A}}^2} \right|_{E_{\mathcal{A}} = E_{\mathcal{A}}^*}$$

We thereby obtain

$$\frac{\delta E_{\mathcal{A}}^*}{E} = \left(\frac{N_{\mathcal{A}} N_{\mathcal{B}}}{N^3} \right)^{1/2} \sim \frac{1}{\sqrt{N}} \sim 10^{-11}$$

for macroscopic systems. Fluctuations in $E_{\mathcal{A}}$ and $E_{\mathcal{B}}$ in such systems are therefore utterly negligible as expected. The mean and most probable values of a sharply peaked probability density very nearly coincide, so $E_{\mathcal{A}}^*$ can alternatively be interpreted as the mean value $\bar{E}_{\mathcal{A}} = \langle E_{\mathcal{A}} \rangle = \int_0^E dE_{\mathcal{A}}\, E_{\mathcal{A}}\, \bar{p}_{\mathcal{A}}(E_{\mathcal{A}}|E)$.

We are now at last in a position to address the correspondence between S and the thermodynamic entropy S_ϑ. This can be done by comparing the thermodynamic and statistical conditions that determine how the total energy E is divided or partitioned between systems $\mathcal{A}$ and $\mathcal{B}$. That partitioning is defined by the value of $E_{\mathcal{A}}^*$ or equivalently $\bar{E}_{\mathcal{A}} = \langle E_{\mathcal{A}} \rangle$. In thermodynamics $\bar{E}_{\mathcal{A}}$ is determined by the condition $T_{\mathcal{A}} = T_{\mathcal{B}}$, whereas in the statistical treatment $E_{\mathcal{A}}^*$ is the value of $E_{\mathcal{A}}$ which maximizes $\bar{p}_{\mathcal{A}}(E_{\mathcal{A}}|E)$. According to Eq. (6.24), the quantities $W_{\mathcal{AB}}(E_{\mathcal{A}}|E)$ and $S_0(E_{\mathcal{A}}|E)$ likewise attain their maxima at $E_{\mathcal{A}} = E_{\mathcal{A}}^*$, so $E_{\mathcal{A}}^*$ can be determined by maximizing $S_0(E_{\mathcal{A}}|E) = S_{\mathcal{A}}(E_{\mathcal{A}}) + S_{\mathcal{B}}(E - E_{\mathcal{A}})$ with respect to $E_{\mathcal{A}}$. Thus we set

$$\frac{\partial S_0(E_{\mathcal{A}}|E)}{\partial E_{\mathcal{A}}} = 0$$

which yields

$$\left.\frac{\partial S_{\mathcal{A}}}{\partial E_{\mathcal{A}}}\right|_{E_{\mathcal{A}}=E_{\mathcal{A}}^*} = \left.\frac{\partial S_{\mathcal{B}}}{\partial E_{\mathcal{B}}}\right|_{E_{\mathcal{B}}=E-E_{\mathcal{A}}^*} \tag{6.28}$$

It is convenient to define

$$\beta(N,V,E) \equiv \left(\frac{\partial S(N,V,E)}{\partial E}\right)_{N,V} \tag{6.29}$$

in terms of which Eq. (6.28) becomes

$$\beta_{\mathcal{A}}(N_{\mathcal{A}}, V_{\mathcal{A}}, E_{\mathcal{A}}^*) = \beta_{\mathcal{B}}(N_{\mathcal{B}}, V_{\mathcal{B}}, E - E_{\mathcal{A}}^*) \tag{6.30}$$

which is a single nonlinear equation in the single unknown variable $E_{\mathcal{A}}^*$, and thereby determines the value of $E_{\mathcal{A}}^*$ as well as the resulting common value of β. Note that $1/\beta$ has units of energy.

Equation (6.30) states that two systems in thermal equilibrium have the same value of β, which strongly suggests a relation between β and temperature. Based on the aforementioned form of $\bar{\omega}(E)$ in simple ideal systems [38], we expect that typically $S \sim \log E$, so that $\beta = \partial S/\partial E \sim 1/E$. But we know from thermodynamics that E increases monotonically with T, and indeed is linearly proportional to T for ideal gases with constant specific heats. Thus $E \sim T$, so $\beta \sim 1/E \sim 1/T$. This inference is qualitatively reinforced by the following observation. Consider the case in which $E_{\mathcal{A}} > E_{\mathcal{A}}^*$ initially, so that energy flows from system $\mathcal{A}$ to system $\mathcal{B}$ when the two are placed in thermal contact. This energy transfer reduces $E_{\mathcal{A}}$ and increases $E_{\mathcal{B}}$, which implies that $\beta_{\mathcal{A}} \sim 1/E_{\mathcal{A}}$ increases while $\beta_{\mathcal{B}} \sim 1/E_{\mathcal{B}}$ decreases. Unlike the energies, however, $\beta_{\mathcal{A}}$ and $\beta_{\mathcal{B}}$ attain the same common value β at equilibrium, so in order for $\beta_{\mathcal{A}}$ to increase to the value β while $\beta_{\mathcal{B}}$ decreases to the same value, their initial values must satisfy $\beta_{\mathcal{A}} < \beta < \beta_{\mathcal{B}}$. According to thermodynamics, energy will flow from system $\mathcal{A}$ to system $\mathcal{B}$ only if their initial temperatures satisfy $T_{\mathcal{A}} > T_{\mathcal{B}}$ or $1/T_{\mathcal{A}} < 1/T_{\mathcal{B}}$, which is indeed the case if $\beta \sim 1/T$. Thus we provisionally presume that $1/\beta = \kappa T$, where κ is an as yet undetermined constant with units of energy divided by temperature. Equation (6.30) then becomes equivalent to the thermodynamic condition $T_{\mathcal{A}} = T_{\mathcal{B}}$ for thermal equilibrium, and is also consistent with energy transfer in the correct direction. Note that κ must be independent of $\mathcal{A}$ or $\mathcal{B}$ in order for $\beta_{\mathcal{A}} = \beta_{\mathcal{B}}$ to imply $T_{\mathcal{A}} = T_{\mathcal{B}}$. This implies that κ is a universal constant, since it does not depend in any way on the nature or constitution of the particular systems $\mathcal{A}$ and $\mathcal{B}$.

Setting $1/\beta = \kappa T$ in Eq. (6.29), we obtain

$$\kappa \left(\frac{\partial S(N, V, E)}{\partial E} \right)_{N,V} = \frac{1}{T} \tag{6.31}$$

which becomes identical to the familiar thermodynamic relation $(\partial S_\vartheta/\partial E)_{N,V} = 1/T$ if we identify κS with the thermodynamic entropy S_ϑ. This observation is highly suggestive, and some authors even seem to consider it definitive, but as yet it merely supports the equivalence of κS and S_ϑ to within an arbitrary function of (N, V). To resolve this ambiguity it is necessary to consider more general interactions between systems $\mathcal{A}$ and $\mathcal{B}$, which is therefore the next task at hand.

6.3 Systems in Thermal and Mechanical Equilibrium

Consider again the same two systems $\mathcal{A}$ and $\mathcal{B}$ as before, but now suppose that when they are placed in contact the common surface between them (which remains thermally conducting and impervious to particles) is no longer fixed but is allowed to move freely, so that the two systems can exchange volume (i.e., expand/contract) as well as energy. Their volumes $V_{\mathcal{A}}$ and $V_{\mathcal{B}}$ are now variable subject to the constraint $V_{\mathcal{A}} + V_{\mathcal{B}} = V$, while $N_{\mathcal{A}}$ and $N_{\mathcal{B}}$ remain fixed as before. Thus $(E_{\mathcal{A}}, E_{\mathcal{B}})$ and $(V_{\mathcal{A}}, V_{\mathcal{B}})$ will change in such a way that the two systems come into simultaneous thermal and mechanical equilibrium, the thermodynamic conditions for which are of course $T_{\mathcal{A}} = T_{\mathcal{B}}$ and $P_{\mathcal{A}} = P_{\mathcal{B}}$.

The description and analysis of simultaneous thermal and mechanical equilibrium requires a generalization of the development of Sect. 6.2 to allow for changes in volume as well as energy. This generalization requires us to address a subtle conceptual issue which pertains to the labeling and enumeration of the microstates, but which does not arise when both N and V are held fixed as they were before. The essential point is that when variations in N and/or V occur, the discrete index k no longer completely specifies the microstate, because the microstates indexed by k also depend parametrically on N and V as discussed in Sect. 6.1. The complete specification of a microstate must therefore include the values of N and V as well as k, and is accordingly defined by the ordered triplet (N, V, k). In the present context N remains fixed and can again be suppressed from the notation, so a particular microstate can be specified or identified by the ordered pair (V, k). Unfortunately, this implies that the microstates are no longer purely discrete; the state index (V, k) is now a hybrid admixture of the discrete parameter k and the continuous parameter V. This complication creates a conundrum in the counting of microstates, as we now proceed to discuss.

To account for simultaneous variations in both E and V, it will be necessary to formally evaluate the number of microstates (V, k) that lie within a finite interval or range of values of V as well as E. However, V is a continuous variable, so the number of values of V within any finite interval is infinite, which implies that the corresponding number of microstates (V, k) is likewise infinite. As discussed in Sect. 2.6, this cannot be tolerated; the number of accessible microstates must be finite in order to obtain a well defined entropy. This dilemma is of essentially the same type as that discussed in connection with continuous states in Sect. 2.6, and can be dealt with in the same way. There it was argued that it is not operationally sensible or meaningful to distinguish between continuous states $\mathbf{x}$ which are arbitrarily close together, and that continuous states should accordingly be described and enumerated in terms of a continuous number density $\rho(\mathbf{x})$ of sensibly distinct states in $\mathbf{x}$-space. In the same spirit, we introduce a continuous number density $g(V)$ of sensibly distinct values of V, which is defined in such a way that $g(V)\,dV$ is the number of such values in the differential increment dV. The integral of $g(V)$ over any finite interval in V then defines the number of sensibly distinct values of V in that interval.

The local mean spacing or increment between sensibly distinct values of V is $\Delta V \equiv 1/g(V)$, which we require to be small in a macroscopic sense (i.e., $\Delta V \ll V$) but large in a microscopic or molecular sense. By this definition, V and V' are considered sensibly distinct only if $|V - V'| \gtrsim \Delta V$. Like E, V is a macroscopic variable which cannot be controlled or measured with infinite precision, so it is natural to regard ΔV as a small uncertainty in V analogous to the small uncertainty ΔE in E. However, ΔV is not a tolerance in quite the same sense as ΔE, because a nonzero value of ΔV is not required to obtain a nonzero value of $W(N, V, E)$ and thereby a well defined entropy. The neighborhood of the volume V is defined by

$$\mathbb{N}(V) \equiv \{V' | V - \tfrac{1}{2}\Delta V < V' < V + \tfrac{1}{2}\Delta V\}$$

Like $\mathbb{N}(E)$ before it, $\mathbb{N}(V)$ corresponds to the neighborhood $\mathbb{N}(\mathbf{a})$ in the general treatment of Chapter 4, so that ΔV corresponds to $v(\mathbf{a})$ and $g(V) = 1/\Delta V$ corresponds to $g(\mathbf{a}) = 1/v(\mathbf{a})$. Since V is another additive macroscopic state variable, ΔV will be taken to be a constant independent of V as discussed in Sect. 4.2.3, which implies that $g(V) = 1/\Delta V$ is likewise independent of V. The number of sensibly distinct values of V in the neighborhood $\mathbb{N}(V)$ is then precisely $\int_{\mathbb{N}(V)} dV'\, g(V') = 1$. Thus $g(V)$ can also be interpreted as the number density of neighborhoods in V, which confirms its correspondence to $g(\mathbf{a})$.

We are now in a position to evaluate the number of microstates (V, k) that lie within finite intervals of V as well as E. As discussed in Sect. 6.1, a differential increment dE in E can be regarded as containing $\bar{\omega}(E|N, V)\, dE = (1/\Delta E)\, W(N, V, E)\, dE$ microstates k for a given fixed value of V, while a differential increment dV in V contains $g(V)\, dV$ sensibly distinct values of V. The number of microstates or ordered pairs (V, k) within the differential area element $dE\, dV$ in the E-V plane is therefore simply the product $\bar{\omega}(E, V|N)\, dE\, dV$, where

$$\bar{\omega}(E, V|N) \equiv g(V)\, \bar{\omega}(E|N, V) = \frac{W(N, V, E)}{\Delta E\, \Delta V}$$

is the density of microstates (V, k) in (E, V)-space. The number of microstates (V, k) within any finite region $\mathbb{R}$ of the E-V plane is then simply $\int_{\mathbb{R}} dE\, dV\, \bar{\omega}(E, V|N)$. In particular, the number of microstates within the small finite area $\Delta E\, \Delta V$ containing the point (E, V) is simply $\bar{\omega}(E, V|N)\, \Delta E\, \Delta V = W(N, V, E)$. The quantity $W(N, V, E)$ thereby acquires a second interpretation as the number of microstates (V, k) within the intervals ΔE and ΔV, in addition to its original interpretation as the number of microstates k in the interval ΔE for fixed V. Of course, this simply reflects the fact that the interval ΔV only contains a single distinct value of V by definition. Note that introducing the uncertainty ΔV therefore does not entail a redefinition of the entropy, which remains $S(N, V, E) = \log W(N, V, E)$ as before.

We can now resume our consideration of the two systems $\mathcal{A}$ and $\mathcal{B}$ and the combined system $\mathcal{AB}$ that results when they are placed in contact. Since $N_{\mathcal{A}}$ and $N_{\mathcal{B}}$ are held fixed, they can be suppressed from the notation as before, so that the various quantities pertaining to systems $\mathcal{A}$ and $\mathcal{B}$, including $W_{\mathcal{A}}(N_{\mathcal{A}}, V_{\mathcal{A}}, E_{\mathcal{A}})$, $S_{\mathcal{A}} = \log W_{\mathcal{A}}$, $W_{\mathcal{B}}(N_{\mathcal{B}}, V_{\mathcal{B}}, E_{\mathcal{B}})$, and $S_{\mathcal{B}} = \log W_{\mathcal{B}}$, can be regarded as functions of $(E_{\mathcal{A}}, V_{\mathcal{A}})$ and $(E_{\mathcal{B}}, V_{\mathcal{B}})$, respectively. Since E and V are held fixed, $E_{\mathcal{B}}$ and $V_{\mathcal{B}}$ are determined by $E_{\mathcal{A}}$ and $V_{\mathcal{A}}$ via the relations $E_{\mathcal{B}} = E - E_{\mathcal{A}}$ and $V_{\mathcal{B}} = V - V_{\mathcal{A}}$, so that only $E_{\mathcal{A}}$ and $V_{\mathcal{A}}$ can vary independently. Thus it is convenient to regard the independent variables of the combined system $\mathcal{AB}$ as $(E_{\mathcal{A}}, V_{\mathcal{A}}, E, V)$ rather than $(E_{\mathcal{A}}, V_{\mathcal{A}}, E_{\mathcal{B}}, V_{\mathcal{B}})$, and moreover to write the former arguments in the form $(E_{\mathcal{A}}, V_{\mathcal{A}}|E, V)$ to emphasize that they denote the functional dependence on $(E_{\mathcal{A}}, V_{\mathcal{A}})$ for given fixed values of E and V. With these conventions, the total number of accessible microstates of the combined system $\mathcal{AB}$ for given values of $E_{\mathcal{A}}$ and $V_{\mathcal{A}}$ within the tolerances ΔE and ΔV is now given by

$$W_{\mathcal{AB}}(E_{\mathcal{A}}, V_{\mathcal{A}}|E, V) \equiv W_{\mathcal{A}}(E_{\mathcal{A}}, V_{\mathcal{A}})\, W_{\mathcal{B}}(E - E_{\mathcal{A}}, V - V_{\mathcal{A}}) \tag{6.32}$$

while the corresponding conditional entropy now becomes

$$\begin{aligned} S_0(E_{\mathcal{A}}, V_{\mathcal{A}}|E, V) &= \log W_{\mathcal{AB}}(E_{\mathcal{A}}, V_{\mathcal{A}}|E, V) \\ &= S_{\mathcal{A}}(E_{\mathcal{A}}, V_{\mathcal{A}}) + S_{\mathcal{B}}(E - E_{\mathcal{A}}, V - V_{\mathcal{A}}) \end{aligned} \tag{6.33}$$

The quantities $W_{\mathcal{AB}}(E_{\mathcal{A}}, V_{\mathcal{A}}|E, V)$ and $S_0(E_{\mathcal{A}}, V_{\mathcal{A}}|E, V)$ can again be interpreted as the number of microstates and the constrained entropy of the combined system $\mathcal{AB}$ for given values of $E_{\mathcal{A}}$ and $V_{\mathcal{A}}$.

According to the second interpretation of $W(N, V, E)$ discussed above, the mean density of microstates of the combined system $\mathcal{AB}$ in $(E_{\mathcal{A}}, V_{\mathcal{A}})$-space is simply

$$\bar{\omega}_{\mathcal{AB}}(E_{\mathcal{A}}, V_{\mathcal{A}}|E, V) \equiv (\Delta E\, \Delta V)^{-1}\, W_{\mathcal{AB}}(E_{\mathcal{A}}, V_{\mathcal{A}}|E, V)$$

When $E_{\mathcal{A}}$ and $V_{\mathcal{A}}$ are allowed to vary freely, the total number of accessible microstates $W(E, V)$ for given fixed values of E and V is therefore given by

$$
\begin{aligned}
W(E,V) &= \int_0^V dV_{\mathcal{A}} \int_0^E dE_{\mathcal{A}}\, \bar{\omega}_{\mathcal{AB}}(E_{\mathcal{A}}, V_{\mathcal{A}}|E,V) \\
&= \frac{1}{\Delta E\, \Delta V} \int_0^V dV_{\mathcal{A}} \int_0^E dE_{\mathcal{A}}\, W_{\mathcal{AB}}(E_{\mathcal{A}}, V_{\mathcal{A}}|E,V) \qquad (6.34)
\end{aligned}
$$

Equation (6.34) now replaces Eq. (6.22) for systems in mechanical as well as thermal equilibrium. The entropy of the unconstrained combined system $\mathcal{AB}$ is then simply $S(E,V) = \log W(E,V)$.

The joint mean probability density $\bar{p}_{\mathcal{A}}(E_{\mathcal{A}}, V_{\mathcal{A}}|E,V)$ in the variables $(E_{\mathcal{A}}, V_{\mathcal{A}})$ for given fixed values of (E,V) is once again simply proportional to the number of microstates $W_{\mathcal{AB}}(E_{\mathcal{A}}, V_{\mathcal{A}}|E,V)$ consistent with $(E_{\mathcal{A}}, V_{\mathcal{A}})$, as given by Eq. (6.32), and of course it must be normalized so that

$$
\int_0^E dE_{\mathcal{A}} \int_0^V dV_{\mathcal{A}}\, \bar{p}_{\mathcal{A}}(E_{\mathcal{A}}, V_{\mathcal{A}}|E,V) = 1
$$

It follows that

$$
\begin{aligned}
\bar{p}_{\mathcal{A}}(E_{\mathcal{A}}, V_{\mathcal{A}}|E,V) &= \frac{W_{\mathcal{AB}}(E_{\mathcal{A}}, V_{\mathcal{A}}|E,V)}{\Delta E\, \Delta V\, W(E,V)} \\
&= \frac{\exp S_0(E_{\mathcal{A}}, V_{\mathcal{A}}|E,V)}{\Delta E\, \Delta V\, \exp S(E,V)} \qquad (6.35)
\end{aligned}
$$

As was the case for $\bar{p}_{\mathcal{A}}(E_{\mathcal{A}}|E)$, $\bar{p}_{\mathcal{A}}(E_{\mathcal{A}}, V_{\mathcal{A}}|E,V)$ is actually independent of ΔE and ΔV, as can be seen by combining Eqs. (6.34) and (6.35). Let $E_{\mathcal{A}}^*$ and $V_{\mathcal{A}}^*$ denote the values of $E_{\mathcal{A}}$ and $V_{\mathcal{A}}$ which maximize $\bar{p}_{\mathcal{A}}(E_{\mathcal{A}}, V_{\mathcal{A}}|E,V)$, and thereby $W_{\mathcal{AB}}(E_{\mathcal{A}}, V_{\mathcal{A}}|E,V)$ and $S_0(E_{\mathcal{A}}, V_{\mathcal{A}}|E,V)$ as well. Equation (6.34) then shows that

$$
\begin{aligned}
W_{\mathcal{AB}}(E_{\mathcal{A}}^*, V_{\mathcal{A}}^*|E,V) &\leq W(E,V) \\
&\leq \left(\frac{EV}{\Delta E\, \Delta V}\right) W_{\mathcal{AB}}(E_{\mathcal{A}}^*, V_{\mathcal{A}}^*|E,V) \qquad (6.36)
\end{aligned}
$$

which combines with Eq. (6.33) to imply

$$\begin{aligned} S_0(E_{\mathcal{A}}^*, V_{\mathcal{A}}^*|E,V) &\leq S(E,V) \\ &\leq S_0(E_{\mathcal{A}}^*, V_{\mathcal{A}}^*|E,V) + \log\left(\frac{EV}{\Delta E\,\Delta V}\right) \end{aligned} \tag{6.37}$$

The logarithmic term involving $E/\Delta E$ and $V/\Delta V$ is clearly utterly negliglble as usual, so Eq. (6.37) reduces to the equality

$$\begin{aligned} S(E,V) &= S_0(E_{\mathcal{A}}^*, V_{\mathcal{A}}^*|E,V) \\ &= S_{\mathcal{A}}(E_{\mathcal{A}}^*, V_{\mathcal{A}}^*) + S_{\mathcal{B}}(E - E_{\mathcal{A}}^*, V - V_{\mathcal{A}}^*) \end{aligned} \tag{6.38}$$

where use has been made of Eq. (6.33).

The implications of Eq. (6.38) are entirely analogous to those discussed following Eq. (6.27), so it seems unnecessary to reiterate and belabor them in the present context. The essential point is that the unconstrained entropy $S(E,V)$ of the combined system $\mathcal{AB}$ in thermal and mechanical equilibrium is equal to the maximum value of the constrained entropy $S_0(E_{\mathcal{A}}, V_{\mathcal{A}}|E,V)$, which moreover is simply the sum of the separate values of $S_{\mathcal{A}}$ and $S_{\mathcal{B}}$ for the isolated systems $\mathcal{A}$ and $\mathcal{B}$ evaluated at the values $(E_{\mathcal{A}}^*, V_{\mathcal{A}}^*)$ and $(E_{\mathcal{B}}^*, V_{\mathcal{B}}^*) = (E - E_{\mathcal{A}}^*, V - V_{\mathcal{A}}^*)$, respectively. The previous argument used to confirm that $\bar{p}_{\mathcal{A}}(E_{\mathcal{A}}|E)$ is sharply peaked about the value $E_{\mathcal{A}} = E_{\mathcal{A}}^*$ can readily be generalized to confirm that $\bar{p}_{\mathcal{A}}(E_{\mathcal{A}}, V_{\mathcal{A}}|E,V)$ is likewise sharply peaked about the values $(E_{\mathcal{A}}^*, V_{\mathcal{A}}^*)$, so that fluctuations in $E_{\mathcal{A}}$ and $V_{\mathcal{A}}$ about those values are negligible. As before, $E_{\mathcal{A}}^*$ and $V_{\mathcal{A}}^*$ can be formally determined by maximizing $\bar{p}_{\mathcal{A}}(E_{\mathcal{A}}, V_{\mathcal{A}}|E,V)$ or $S_0(E_{\mathcal{A}}, V_{\mathcal{A}}|E,V)$ with respect to $E_{\mathcal{A}}$ and $V_{\mathcal{A}}$. We therefore set

$$\left(\frac{\partial S_0(E_{\mathcal{A}}, V_{\mathcal{A}}|E,V)}{\partial E_{\mathcal{A}}}\right)_{V_{\mathcal{A}}} = \left(\frac{\partial S_0(E_{\mathcal{A}}, V_{\mathcal{A}}|E,V)}{\partial V_{\mathcal{A}}}\right)_{E_{\mathcal{A}}} = 0$$

which combine with Eq. (6.33) to yield

$$\beta_{\mathcal{A}}(N_{\mathcal{A}}, V_{\mathcal{A}}^*, E_{\mathcal{A}}^*) = \beta_{\mathcal{B}}(N_{\mathcal{B}}, V - V_{\mathcal{A}}^*, E - E_{\mathcal{A}}^*) \tag{6.39}$$

$$\varphi_{\mathcal{A}}(N_{\mathcal{A}}, V_{\mathcal{A}}^*, E_{\mathcal{A}}^*) = \varphi_{\mathcal{B}}(N_{\mathcal{B}}, V - V_{\mathcal{A}}^*, E - E_{\mathcal{A}}^*) \tag{6.40}$$

where $\beta(N, V, E)$ is defined by Eq. (6.29), and

$$\varphi(N, V, E) \equiv \left(\frac{\partial S(N, V, E)}{\partial V}\right)_{N,E} \tag{6.41}$$

Equations (6.39) and (6.40) constitute two simultaneous nonlinear equations in the two unknown variables $E_{\mathcal{A}}^*$ and $V_{\mathcal{A}}^*$, and thereby determine the values of those variables as well as the resulting common values of β and φ.

Equations (6.39) and (6.40) state that two systems in thermal and mechanical equilibrium have the same values of both β and φ. The equality of β was previously interpreted in Sect. 6.2 as the statistical analog of the thermodynamic condition $T_{\mathcal{A}} = T_{\mathcal{B}}$ for thermal equilibrium, which led to the tentative identifications of $\kappa\beta$ with $1/T$ and κS with the thermodynamic entropy S_ϑ. It is therefore natural to further presume that the equality of φ is the statistical analog of the thermodynamic condition $P_{\mathcal{A}} = P_{\mathcal{B}}$ for mechanical equilibrium. Indeed, Eq. (6.41) becomes identical to the familiar thermodynamic relation $(\partial S_\vartheta/\partial V)_{N,E} = P/T$ if we continue to identify κS with S_ϑ and also identify $\kappa\varphi$ with P/T. This further reinforces the identification of κS with S_ϑ, which has now been shown to be consistent to within an arbitrary function of N. This final remaining ambiguity is removed in the next section.

6.4 Systems in Thermal and Diffusional Equilibrium

Consider again two initially isolated systems $\mathcal{A}$ and $\mathcal{B}$ which are placed in contact so that they interact with each other by sharing a common surface, which is now assumed to be rigid and immovable but is both thermally conducting and permeable or porous to particles. The two systems can then exchange energy and particles, but not volume. In contrast to the previous cases considered in Sects. 6.2 and 6.3, we must now require that systems $\mathcal{A}$ and $\mathcal{B}$ are both composed of the same type or species of particles, so that they remain pure single-component systems even after they exchange particles. Thus $N_{\mathcal{A}}$ and $N_{\mathcal{B}}$ are now variable subject to the constraint $N_{\mathcal{A}} + N_{\mathcal{B}} = N$, whereas $V_{\mathcal{A}}$ and $V_{\mathcal{B}}$ remain fixed. When systems $\mathcal{A}$ and $\mathcal{B}$ are placed in contact, $(E_{\mathcal{A}}, E_{\mathcal{B}})$ and $(N_{\mathcal{A}}, N_{\mathcal{B}})$ will now change in such a way that the two systems come into simultaneous thermal and diffusional equilibrium with each other. The thermodynamic conditions for those equilibria are of course $T_{\mathcal{A}} = T_{\mathcal{B}}$ and $\mu_{\mathcal{A}} = \mu_{\mathcal{B}}$, where μ is the chemical potential.

The present treatment is largely isomorphic to that of Sect. 6.3 with the roles of $(V_{\mathcal{A}}, V_{\mathcal{B}})$ and $(N_{\mathcal{A}}, N_{\mathcal{B}})$ interchanged, but is rather simpler since $N_{\mathcal{A}}$ and $N_{\mathcal{B}}$ are already discrete variables, so the complications of dealing with a continuum of microstates do not arise. Since the volumes $V_{\mathcal{A}}$ and $V_{\mathcal{B}}$ are now held fixed, they can be suppressed from the notation. The microstates of each system can then be specified or identified by discrete ordered pairs of the form (N, k). Since N is already discrete, there is no need to introduce a finite interval ΔN in order to evaluate the number of microstates that lie within a range of its values, which merely requires a summation over N. The various quantities pertaining to systems $\mathcal{A}$ and $\mathcal{B}$ can now be regarded as functions of $(E_{\mathcal{A}}, N_{\mathcal{A}})$ and $(E_{\mathcal{B}}, N_{\mathcal{B}})$, respectively. Since E and N are held fixed, $E_{\mathcal{B}}$ and $N_{\mathcal{B}}$ are determined by $E_{\mathcal{A}}$ and $N_{\mathcal{A}}$ via the relations $E_{\mathcal{B}} = E - E_{\mathcal{A}}$ and $N_{\mathcal{B}} = N - N_{\mathcal{A}}$, so that only $E_{\mathcal{A}}$ and $N_{\mathcal{A}}$ can vary independently. Thus it is convenient to regard the independent variables of the combined system $\mathcal{AB}$ as $(E_{\mathcal{A}}, N_{\mathcal{A}}, E, N)$ rather than $(E_{\mathcal{A}}, N_{\mathcal{A}}, E_{\mathcal{B}}, N_{\mathcal{B}})$, and moreover to write the former arguments in the form $(E_{\mathcal{A}}, N_{\mathcal{A}}|E, N)$ to emphasize that they denote the functional dependence on $(E_{\mathcal{A}}, N_{\mathcal{A}})$ for given fixed values of E and N. With these conventions, the total number of accessible microstates of the combined system $\mathcal{AB}$ for given values of $E_{\mathcal{A}}$ and $N_{\mathcal{A}}$ is now given by

$$W_{\mathcal{AB}}(E_{\mathcal{A}}, N_{\mathcal{A}}|E, N) \equiv W_{\mathcal{A}}(E_{\mathcal{A}}, N_{\mathcal{A}})\, W_{\mathcal{B}}(E - E_{\mathcal{A}}, N - N_{\mathcal{A}}) \quad (6.42)$$

while the corresponding conditional entropy becomes

$$\begin{aligned} S_0(E_{\mathcal{A}}, N_{\mathcal{A}}|E, N) &= \log W_{\mathcal{AB}}(E_{\mathcal{A}}, N_{\mathcal{A}}|E, N) \\ &= S_{\mathcal{A}}(E_{\mathcal{A}}, N_{\mathcal{A}}) + S_{\mathcal{B}}(E - E_{\mathcal{A}}, N - N_{\mathcal{A}}) \end{aligned} \tag{6.43}$$

The quantities $W_{\mathcal{AB}}(E_{\mathcal{A}}, N_{\mathcal{A}}|E, N)$ and $S_0(E_{\mathcal{A}}, N_{\mathcal{A}}|E, N)$ can again be interpreted as the number of microstates and the constrained entropy of the combined system $\mathcal{AB}$ for given values of $E_{\mathcal{A}}$ and $N_{\mathcal{A}}$.

For each discrete value of $N_{\mathcal{A}}$, the mean density of microstates of the combined system $\mathcal{AB}$ in $E_{\mathcal{A}}$ is simply the number of such microstates per unit increment in $E_{\mathcal{A}}$, which is given by

$$\bar{\omega}_{\mathcal{AB}}(E_{\mathcal{A}}, N_{\mathcal{A}}|E, N) \equiv (1/\Delta E)\, W_{\mathcal{AB}}(E_{\mathcal{A}}, N_{\mathcal{A}}|E, N)$$

When $E_{\mathcal{A}}$ and $N_{\mathcal{A}}$ are allowed to vary freely, the total number $W(E, N)$ of accessible microstates of the combined system $\mathcal{AB}$ for given fixed values of E and N is therefore given by

$$\begin{aligned} W(E, N) &= \sum_{N_{\mathcal{A}}=0}^{N} \int_0^E dE_{\mathcal{A}}\, \bar{\omega}_{\mathcal{AB}}(E_{\mathcal{A}}, N_{\mathcal{A}}|E, N) \\ &= \frac{1}{\Delta E} \sum_{N_{\mathcal{A}}=0}^{N} \int_0^E dE_{\mathcal{A}}\, W_{\mathcal{AB}}(E_{\mathcal{A}}, N_{\mathcal{A}}|E, N) \end{aligned} \tag{6.44}$$

Equation (6.44) now replaces Eq. (6.22) for systems in diffusional as well as thermal equilibrium. The entropy of the unconstrained combined system $\mathcal{AB}$ is then simply $S(E, N) = \log W(E, N)$.

The joint probability density $\bar{p}_{\mathcal{A}}(E_{\mathcal{A}}, N_{\mathcal{A}}|E, N)$ in the variables $(E_{\mathcal{A}}, N_{\mathcal{A}})$ for given fixed values of (E, N) is once again simply proportional to the number of microstates $W_{\mathcal{AB}}(E_{\mathcal{A}}, N_{\mathcal{A}}|E, N)$ consistent with $(E_{\mathcal{A}}, N_{\mathcal{A}})$, as given by Eq. (6.42), and of course it must be normalized so that

$$\sum_{N_{\mathcal{A}}=0}^{N} \int_0^E dE_{\mathcal{A}}\, \bar{p}_{\mathcal{A}}(E_{\mathcal{A}}, N_{\mathcal{A}}|E, N) = 1$$

It follows that

$$\begin{aligned} \bar{p}_{\mathcal{A}}(E_{\mathcal{A}}, N_{\mathcal{A}}|E, N) &= \frac{W_{\mathcal{AB}}(E_{\mathcal{A}}, N_{\mathcal{A}}|E, N)}{\Delta E\, W(E, N)} \\ &= \frac{\exp S_0(E_{\mathcal{A}}, N_{\mathcal{A}}|E, N)}{\Delta E\ \exp S(E, N)} \end{aligned} \tag{6.45}$$

As was the case for $\bar{p}_{\mathcal{A}}(E_{\mathcal{A}}|E)$, $\bar{p}_{\mathcal{A}}(E_{\mathcal{A}}, N_{\mathcal{A}}|E, N)$ is independent of ΔE, as can be seen by combining Eqs. (6.44) and (6.45). Let $E_{\mathcal{A}}^*$ and $N_{\mathcal{A}}^*$ denote the values of $E_{\mathcal{A}}$ and $N_{\mathcal{A}}$ for which $\bar{p}_{\mathcal{A}}(E_{\mathcal{A}}, N_{\mathcal{A}}|E, N)$ is a maximum. According to Eq. (6.45), $W_{\mathcal{AB}}(E_{\mathcal{A}}, N_{\mathcal{A}}|E, N)$ and $S_0(E_{\mathcal{A}}, N_{\mathcal{A}}|E, N)$ attain their maxima at those same values. Equation (6.44) then shows that $W(E, N)$ satisfies the inequalities

$$\begin{aligned} W_{\mathcal{AB}}(E_{\mathcal{A}}^*, N_{\mathcal{A}}^*|E, V) &\leq W(E, N) \\ &\leq (NE/\Delta E)\, W_{\mathcal{AB}}(E_{\mathcal{A}}^*, N_{\mathcal{A}}^*|E, N) \end{aligned} \tag{6.46}$$

which combine with Eq. (6.43) to imply

$$\begin{aligned} S_0(E_{\mathcal{A}}^*, N_{\mathcal{A}}^*|E, N) &\leq S(E, N) \\ &\leq S_0(E_{\mathcal{A}}^*, N_{\mathcal{A}}^*|E, N) + \log\left(\frac{NE}{\Delta E}\right) \end{aligned} \tag{6.47}$$

The term $\log(NE/\Delta E)$ is negligible, because $\log(E/\Delta E)$ is negligible as usual, $\log N \ll N$ for large N, and $S \sim N$ as previously remarked. Equation (6.47) therefore reduces to the equality

$$\begin{aligned} S(E, N) &= S_0(E_{\mathcal{A}}^*, N_{\mathcal{A}}^*|E, N) \\ &= S_{\mathcal{A}}(E_{\mathcal{A}}^*, N_{\mathcal{A}}^*) + S_{\mathcal{B}}(E - E_{\mathcal{A}}^*, N - N_{\mathcal{A}}^*) \end{aligned} \tag{6.48}$$

where use has been made of Eq. (6.43).

The implications of Eq. (6.48) are again entirely analogous to those discussed following Eq. (6.27), which by now should require no further explanation. All that remains is to formally determine the values of $E_{\mathcal{A}}^*$ and $N_{\mathcal{A}}^*$ that maximize $\bar{p}_{\mathcal{A}}(E_{\mathcal{A}}, N_{\mathcal{A}}|E, N)$, or equivalently

$S_0(E_{\mathcal{A}}, N_{\mathcal{A}}|E, N)$. Since $N^*_{\mathcal{A}}$ is very large, $N_{\mathcal{A}}$ can be treated as a continuous variable for this purpose. Thus we set

$$\left(\frac{\partial S_0(E_{\mathcal{A}}, N_{\mathcal{A}}|E, N)}{\partial E_{\mathcal{A}}}\right)_{N_{\mathcal{A}}} = \left(\frac{\partial S_0(E_{\mathcal{A}}, N_{\mathcal{A}}|E, N)}{\partial N_{\mathcal{A}}}\right)_{E_{\mathcal{A}}} = 0$$

which combine with Eq. (6.43) to yield

$$\beta_{\mathcal{A}}(N^*_{\mathcal{A}}, V_{\mathcal{A}}, E^*_{\mathcal{A}}) = \beta_{\mathcal{B}}(N - N^*_{\mathcal{A}}, V_{\mathcal{B}}, E - E^*_{\mathcal{A}}) \tag{6.49}$$

$$\eta_{\mathcal{A}}(N^*_{\mathcal{A}}, V_{\mathcal{A}}, E^*_{\mathcal{A}}) = \eta_{\mathcal{B}}(N - N^*_{\mathcal{A}}, V_{\mathcal{B}}, E - E^*_{\mathcal{A}}) \tag{6.50}$$

where $\beta(N, V, E)$ is still defined by Eq. (6.29), and

$$\eta(N, V, E) \equiv \left(\frac{\partial S(N, V, E)}{\partial N}\right)_{V,E} \tag{6.51}$$

Equations (6.49) and (6.50) constitute two simultaneous nonlinear equations in the two unknown variables $E^*_{\mathcal{A}}$ and $N^*_{\mathcal{A}}$, and thereby determine the values of those variables as well as the resulting common values of β and η. However, the fact that systems $\mathcal{A}$ and $\mathcal{B}$ are now composed of the same type or species of particles implies that their entropies $S_{\mathcal{A}}$ and $S_{\mathcal{B}}$ are the same function S of their respective arguments; i.e., $S_{\mathcal{A}} = S(N_{\mathcal{A}}, V_{\mathcal{A}}, E_{\mathcal{A}})$ and $S_{\mathcal{B}} = S(N_{\mathcal{B}}, V_{\mathcal{B}}, E_{\mathcal{B}})$. It follows that $\beta_{\mathcal{A}}$ and $\beta_{\mathcal{B}}$ are now likewise the same function of their respective arguments, and so are $\eta_{\mathcal{A}}$ and $\eta_{\mathcal{B}}$, so that the subscripts $\mathcal{A}$ and $\mathcal{B}$ can be removed from the functions $\beta(N, V, E)$ and $\eta(N, V, E)$. Equations (6.49) and (6.50) thereby reduce to the simpler forms

$$\beta(N^*_{\mathcal{A}}, V_{\mathcal{A}}, E^*_{\mathcal{A}}) = \beta(N - N^*_{\mathcal{A}}, V_{\mathcal{B}}, E - E^*_{\mathcal{A}}) \tag{6.52}$$

$$\eta(N^*_{\mathcal{A}}, V_{\mathcal{A}}, E^*_{\mathcal{A}}) = \eta(N - N^*_{\mathcal{A}}, V_{\mathcal{B}}, E - E^*_{\mathcal{A}}) \tag{6.53}$$

Equations (6.52) and (6.53) state that two systems in thermal and diffusional equilibrium have the same values of both β and η. The analyses of thermal and mechanical equilibrium in Sects. 6.2 and 6.3 resulted in the identifications of $\kappa\beta$ with $1/T$ and $\kappa\varphi$ with P/T. In the same way, the present analysis of diffusional equilibrium strongly suggests that the condition $\eta_A = \eta_B$ is the statistical analog of the thermodynamic condition $\mu_A = \mu_B$. Indeed, Eq. (6.51) becomes identical to the familiar thermodynamic relation $(\partial S_\vartheta/\partial N)_{V,E} = -\mu/T$ if we continue to identify κS with the thermodynamic entropy S_ϑ, and further identify $\kappa\eta$ with $-\mu/T$. This final identification completes the process of relating the statistical quantities (β, φ, η) to thermodynamic variables. The resulting relations

$$\beta = \frac{1}{\kappa T} \quad ; \quad \varphi = \frac{P}{\kappa T} \quad ; \quad \eta = -\frac{\mu}{\kappa T} \tag{6.54}$$

then combine with the purely statistical identity

$$dS = \beta\, dE + \varphi\, dV + \eta\, dN$$

to yield

$$\kappa T dS = dE + P\, dV - \mu\, dN \tag{6.55}$$

Comparison of Eq. (6.55) with the fundamental thermodynamic identity

$$T\, dS_\vartheta = dE + P\, dV - \mu\, dN \tag{6.56}$$

then shows that interpreting κS as S_ϑ is consistent to within an arbitrary additive constant independent of (N, V, E). But the thermodynamic entropy is itself indeterminate to within an arbitrary additive constant, which can therefore be chosen in such a way as to make S_ϑ identical to κS. We can therefore regard the relation $S_\vartheta = \kappa S$ as having been established beyond a reasonable doubt. However, the value of κ is as yet unknown, and is obviously required in order to compute the actual numerical value of S_ϑ from that of S. The statistical interpretation of thermodynamics therefore remains incomplete until κ has been determined. A provisional determination of κ is carried out in the next section, but it will not be definitively confirmed until Chapter 8.

The reader might well inquire why we did not establish the relations involving equilibration of pressure and chemical potential by respectively considering systems which are allowed to exchange only volume or particles, but not energy. The reason is that it is difficult to imagine a moveable partition which does not exchange energy as well as volume, because volume changes do $P\,dV$ work. Similarly, it is difficult to imagine a fixed porous partition which does not exchange energy as well as particles, since particles carry energy with them when they pass through the partition. Conversely, the reader might equally well inquire why we did not unify the development of Sects. 6.2 – 6.4 by performing a single analysis in which the partition separating systems $\mathcal{A}$ and $\mathcal{B}$ is presumed to be simultaneously thermally conducting, moveable, and porous to particles, thereby allowing the two systems to exchange energy, volume, and particles simultaneously. Some authors indeed do this, but such a procedure is fallacious. The difficulty is that a partition which does not constrain anything and through which everything may pass freely is essentially equivalent to no partition at all and is effectively nonexistent. This is a degenerate situation in which the final equilibrium state is ambiguous, because it leaves the final equilibrated values of $N_{\mathcal{A}}$, $V_{\mathcal{A}}$, and $E_{\mathcal{A}}$ undetermined to within a constant factor. This degeneracy is further reflected in the fact that there are only two independent intensive thermodynamic variables in the thermodynamic limit, so that only two of the equilibrium conditions $T_{\mathcal{A}} = T_{\mathcal{B}}$, $P_{\mathcal{A}} = P_{\mathcal{B}}$, and $\mu_{\mathcal{A}} = \mu_{\mathcal{B}}$ are independent while the third is redundant, which again leaves the values of $N_{\mathcal{A}}$, $V_{\mathcal{A}}$, and $E_{\mathcal{A}}$ undetermined to within a constant factor. This ambiguity is related to the fact that a proper set of independent thermodynamic state variables must include at least one extensive variable to specify the size of the system [39–42].

6.5 The Value of κ and the Form of $W(N,V,E)$

Since κ is a universal constant, the obvious way to determine it is to independently evaluate both S_{ϑ} and S for some particular physical system which is sufficiently simple for analytical results to be obtained, and then infer the value of κ by comparing those results. The usual and obvious choice for such a system is an ideal gas. However, it is sufficient and much simpler to evaluate and compare the derivatives of S_ϑ and S that appear in Eq. (6.54). In particular, the pressure of an ideal gas is given by $P = (N/V)k_BT$, where k_B is Boltzmann's constant. This combines with Eq. (6.54) to yield

$$\kappa\varphi = (N/V)\,k_B \tag{6.57}$$

In order to evaluate $\varphi = (\partial S/\partial V)_{N,E}$, the dependence of $S = \log W(N, V, E)$ on V must be determined for an ideal gas. This in turn requires identification and enumeration of the microstates of a system of N identical noninteracting particles, each of which possesses its own single-particle microstates. This is a nontrivial task which is fundamental to statistical mechanics, and it requires a careful explication which will be developed in Chapter 8. For the present we simply resort to a heuristic argument, the validity of which will be confirmed in Chapter 8. The argument is based on the intuitive expectation that the number of microstates accessible to each individual particle in an ideal gas should be proportional to V for fixed (N, E). The number of microstates for the entire gas should then be proportional to V^N, with a coefficient of proportionality that depends on (N, E), so that $W(N, V, E)$ is of the form $W(N, V, E) = V^N C(N, E)$ [43]. (As will be discussed in Chapter 8, V^N should actually be divided by $N!$ to account for the indistinguishability of identical particles, but the factor of $1/N!$ can be absorbed into $C(N, E)$ and is therefore immaterial for present purposes.) It follows at once that $\varphi = N/V$, which combines with Eq. (6.57) to imply that $\kappa = k_B$; i.e., κ is simply Boltzmann's constant. It would be wasteful of space to rewrite all the previous equations involving κ by replacing it with k_B therein, so it will henceforth simply be understood that κ and k_B are identical and the two symbols are synonymous.

The functional form of $S_\vartheta(N, V, E)$ for a monatomic ideal gas with constant specific heats is already known from thermodynamics, and is given by

$$S_\vartheta = Nk_B \log \left[C_0 \left(\frac{V}{N} \right) \left(\frac{E}{N} \right)^{3/2} \right]$$

where C_0 is a constant independent of (N, V, E). The relation $S_\vartheta = k_B S = k_B \log W$ enables us to infer the asymptotic dependence of $W(N, V, E)$ on V and E to logarithmic accuracy by reverse-engineering of S_ϑ, so to speak. We thereby obtain

$$W(N, V, E) \sim V^N E^{3N/2}$$

which confirms and combines the previous assertions $W \sim E^{3N/2}$ and $W \sim V^N$. Although this expression is strictly valid only for monatomic ideal gases, it is useful to keep it in mind as a prototypical example for intuitive purposes, since it explicitly exhibits the steep exponential dependence of W on E and V.

6.6 Thermodynamic Fluctuations

The probability densities $\bar{p}_{\mathcal{A}}(E_{\mathcal{A}}|E)$, $\bar{p}_{\mathcal{A}}(E_{\mathcal{A}}, V_{\mathcal{A}}|E, V)$, and $\bar{p}_{\mathcal{A}}(E_{\mathcal{A}}, N_{\mathcal{A}}|E, N)$ given in Eqs. (6.24), (6.35), and (6.45) describe fluctuations in the quantities $E_{\mathcal{A}}$, $V_{\mathcal{A}}$, and/or $N_{\mathcal{A}}$ in a macroscopic system $\mathcal{A}$ which is in thermal, mechanical, and/or diffusional equilibrium with another macroscopic system $\mathcal{B}$. As previously discussed, those probability densities can be presumed to be sharply peaked about the values $E_{\mathcal{A}}^*$, $V_{\mathcal{A}}^*$, and/or $N_{\mathcal{A}}^*$. They can then be approximated by Gaussian distributions as described in Sect. 4.2.2, from which relatively simple expressions can be derived for the mean square deviations of those quantities from their mean values. The probability densities are also susceptible to a second type of approximation in situations where system $\mathcal{B}$ is much larger than system $\mathcal{A}$, and thereby serves as an essentially infinite bath or reservoir of heat, volume, and/or particles. These two types of simplifying approximations to $\bar{p}_{\mathcal{A}}(E_{\mathcal{A}}|E)$ are developed in Sects. 6.6.1 and 6.6.2, respectively. These results apply to the most important case in which systems $\mathcal{A}$ and $\mathcal{B}$ are in thermal equilibrium, but not mechanical or diffusional equilibrium. The corresponding treatment of the simultaneous fluctuations in $E_{\mathcal{A}}$ and $V_{\mathcal{A}}$ or $N_{\mathcal{A}}$ described by $\bar{p}_{\mathcal{A}}(E_{\mathcal{A}}, V_{\mathcal{A}}|E, V)$ or $\bar{p}_{\mathcal{A}}(E_{\mathcal{A}}, N_{\mathcal{A}}|E, N)$ is more tedious algebraically but is entirely similar and involves no additional concepts, so it will be left as an exercise for interested readers. Generalized results of the same type can readily be derived for systems with additional thermodynamic variables by applying the formalism developed in Sect. 4.2 to composite systems $\mathcal{AB}$ and identifying the state variables **A** with the properties of system $\mathcal{A}$ alone, as discussed in Sect. 4.2.3. Such results are most frequently encountered in the simplified forms they assume when one of the systems is much larger than the other.

6.6.1 Fluctuations About Mean Values

We begin by rewriting Eq. (6.24) in the more convenient equivalent form

$$\bar{p}_{\mathcal{A}}(E_{\mathcal{A}}|E) = C_{\mathcal{A}}(E) \exp S_0(E_{\mathcal{A}}|E) \tag{6.58}$$

where the normalization coefficient $C_{\mathcal{A}}(E)$ is given by

$$C_{\mathcal{A}}(E) = \left[\int_0^E dE_{\mathcal{A}} \, \exp S_0(E_{\mathcal{A}}|E) \right]^{-1} \tag{6.59}$$

and is understood that any approximations to $S_0(E_{\mathcal{A}}|E)$ in Eq. (6.58) must be made in Eq. (6.59) as well in order to preserve normalization. Since $\bar{p}_{\mathcal{A}}(E_{\mathcal{A}}|E)$ is sharply peaked, it is negligible except when $E_{\mathcal{A}} \cong E_{\mathcal{A}}^*$. This implies that an approximation to $S_0(E_{\mathcal{A}}|E)$ only needs to be accurate for $|E_{\mathcal{A}} - E_{\mathcal{A}}^*| \ll E_{\mathcal{A}}^*$, so that $S_0(E_{\mathcal{A}}|E)$ can be approximated by the first few terms in a Taylor series expansion about the value $E_{\mathcal{A}} = E_{\mathcal{A}}^*$. The linear term in the expansion vanishes because $S_0(E_{\mathcal{A}}|E)$ is a maximum at $E_{\mathcal{A}} = E_{\mathcal{A}}^*$, so it is necessary to retain the quadratic term to obtain a useful approximation. We thereby obtain

$$S_0(E_{\mathcal{A}}|E) \cong S(E) - \tfrac{1}{2} D_{\mathcal{A}} (E_{\mathcal{A}} - E_{\mathcal{A}}^*)^2 \tag{6.60}$$

where use has been made of Eq. (6.27), and

$$D_{\mathcal{A}} \equiv -\left.\frac{\partial^2 S_0(E_{\mathcal{A}}|E)}{\partial E_{\mathcal{A}}^2}\right|_{E_{\mathcal{A}}=E_{\mathcal{A}}^*} \tag{6.61}$$

Combining Eqs. (6.58)–(6.60), we obtain the Gaussian distribution

$$\bar{p}_{\mathcal{A}}(E_{\mathcal{A}}|E) \cong \left(\frac{D_{\mathcal{A}}}{2\pi}\right)^{1/2} \exp\{-\tfrac{1}{2}D_{\mathcal{A}}(E_{\mathcal{A}} - E_{\mathcal{A}}^*)^2\} \tag{6.62}$$

where the integration limits in Eq. (6.59) have been extended to $\pm\infty$ with negligible error. Combining Eqs. (6.21) and (6.61), we obtain, after a little algebra,

$$D_{\mathcal{A}} = \frac{1}{\kappa T^2}\left(\frac{1}{C_V^{\mathcal{A}}} + \frac{1}{C_V^{\mathcal{B}}}\right) \tag{6.63}$$

where T is the common temperature of systems $\mathcal{A}$ and $\mathcal{B}$, $C_V^{\mathcal{A}}$ and $C_V^{\mathcal{B}}$ are their respective heat capacities at constant volume (i.e., $C_V \equiv (\partial E/\partial T)_{N,V}$), and use has been made of Eqs. (6.29), (6.30), and (6.54). The variance of the fluctuations in $E_{\mathcal{A}}$ is defined by

$$\delta E_{\mathcal{A}}^2 \equiv \langle (E_{\mathcal{A}} - E_{\mathcal{A}}^*)^2 \rangle \equiv \int dE_{\mathcal{A}}\, \bar{p}_{\mathcal{A}}(E_{\mathcal{A}}|E)\,(E_{\mathcal{A}} - E_{\mathcal{A}}^*)^2 \tag{6.64}$$

and its well known value for the Gaussian distribution of Eq. (6.62) is simply $\delta E_{\mathcal{A}}^2 = 1/D_{\mathcal{A}}$, so

$$\delta E_{\mathcal{A}}^2 = \kappa T^2 \left(\frac{C_V^{\mathcal{A}} C_V^{\mathcal{B}}}{C_V^{\mathcal{A}} + C_V^{\mathcal{B}}}\right) \tag{6.65}$$

For ideal gases $C_V^{\mathcal{A}} \sim \kappa N_{\mathcal{A}}$, $C_V^{\mathcal{B}} \sim \kappa N_{\mathcal{B}}$, and $E \sim N\kappa T$, where $N = N_{\mathcal{A}} + N_{\mathcal{B}}$. Equation (6.65) then reduces to $\delta E_{\mathcal{A}}^2 \sim (\kappa T)^2(N_{\mathcal{A}}N_{\mathcal{B}}/N)$, so $\delta E_{\mathcal{A}}/E \sim (N_{\mathcal{A}}N_{\mathcal{B}}/N^3)^{1/2} \sim 1/\sqrt{N}$, in agreement with our previous estimate in Sect. 6.2.

6.6.2 Fluctuations in Systems Coupled to Reservoirs

We now consider situations in which system $\mathcal{B}$ is much larger than system $\mathcal{A}$, so that $\bar{E}_{\mathcal{A}} \ll E$. Since $E_{\mathcal{A}}$ itself is not held fixed, it can in principle have any value between zero and E, but its probability density $\bar{p}_{\mathcal{A}}(E_{\mathcal{A}}|E)$ differs significantly from zero only when $E_{\mathcal{A}} \sim \bar{E}_{\mathcal{A}}$. Thus $\bar{p}_{\mathcal{A}}(E_{\mathcal{A}}|E)$ can be accurately approximated by presuming that $E_{\mathcal{A}} = E - E_{\mathcal{B}} \ll E$, so that $S_{\mathcal{B}}(E_{\mathcal{B}}) = S_{\mathcal{B}}(E - E_{\mathcal{A}}) \cong S_{\mathcal{B}}(E)$. Thus $S_{\mathcal{B}}(E)$ can be accurately approximated by

$$S_{\mathcal{B}}(E) = S_{\mathcal{B}}(E - E_{\mathcal{A}}) + \beta_{\mathcal{B}}(E - E_{\mathcal{A}})E_{\mathcal{A}} \qquad (6.66)$$

where $\beta_{\mathcal{B}}(E_{\mathcal{B}})$ is defined by Eq. (6.29). But according to Eq. (6.30), $\beta_{\mathcal{B}}(E-E_{\mathcal{A}})$ is just the common value of $\beta = 1/(\kappa T)$ at which systems $\mathcal{A}$ and $\mathcal{B}$ come into thermal equilibrium. Equations (6.21) and (6.66) then combine to yield

$$S_0(E_{\mathcal{A}}|E) = S_{\mathcal{A}}(E_{\mathcal{A}}) + S_{\mathcal{B}}(E) - \beta E_{\mathcal{A}} \qquad (6.67)$$

which is equivalent to

$$S_0(E_{\mathcal{A}}|E) = S_{\mathcal{B}}(E) - \beta F_{\mathcal{A}}(E_{\mathcal{A}}) \qquad (6.68)$$

where $F_{\mathcal{A}}(E_{\mathcal{A}}) \equiv E_{\mathcal{A}} - \kappa T S_{\mathcal{A}}(E_{\mathcal{A}})$ is the Helmholtz free energy of system $\mathcal{A}$. Combining Eqs. (6.58), (6.59), and (6.68), we obtain

$$\bar{p}_{\mathcal{A}}(E_{\mathcal{A}}|E) = \frac{\exp\{-\beta F_{\mathcal{A}}(E_{\mathcal{A}})\}}{\int dE'_{\mathcal{A}} \exp\{-\beta F_{\mathcal{A}}(E'_{\mathcal{A}})\}} \qquad (6.69)$$

where the integral extends over the range $0 \leq E'_{\mathcal{A}} \leq E$, although the upper limit can be extended to ∞ with negligible error. Equation (6.69) shows that $F_{\mathcal{A}}(E_{\mathcal{A}})$ must be a minimum at $E_{\mathcal{A}} = E^*_{\mathcal{A}}$, where $\bar{p}_{\mathcal{A}}(E_{\mathcal{A}}|E)$ is a maximum. This can be directly verified by setting $\partial F_{\mathcal{A}}(E_{\mathcal{A}})/\partial E_{\mathcal{A}} = 1 - \kappa T \beta_{\mathcal{A}}(E_{\mathcal{A}}) = 0$. The minimum value of $F_{\mathcal{A}}(E_{\mathcal{A}})$ therefore occurs when $\beta_{\mathcal{A}}(E_{\mathcal{A}}) = 1/(\kappa T)$, the solution of which is indeed $E_{\mathcal{A}} = E^*_{\mathcal{A}}$ according to Eq. (6.30). This confirms the familiar thermodynamic condition that the Helmholtz free energy $F_{\mathcal{A}}$ is a minimum for a system of fixed (N, V) in thermal equilibrium with a heat bath at temperature T.

From a statistical perspective, the presence of $-\beta F_{\mathcal{A}}$ rather than $-\beta E_{\mathcal{A}}$ in the exponential in Eq. (6.69) merely reflects and accounts for the fact that the value of $E_{\mathcal{A}}$ is common to a group of microstates of system $\mathcal{A}$ rather than a single microstate. The number of microstates in that group is $W_{\mathcal{A}}(E_{\mathcal{A}}) = \exp S_{\mathcal{A}}(E_{\mathcal{A}}) = \bar{\omega}_{\mathcal{A}}(E_{\mathcal{A}})\Delta E$, so $\beta F_{\mathcal{A}} = \beta E_{\mathcal{A}} - \log W_{\mathcal{A}}$. Equation (6.69) can therefore be rewritten in the arguably more fundamental form

$$\bar{p}_{\mathcal{A}}(E_{\mathcal{A}}|E) = (1/\bar{Z}_{\mathcal{A}})\,\bar{\omega}_{\mathcal{A}}(E_{\mathcal{A}})\exp(-\beta E_{\mathcal{A}}) \qquad (6.70)$$

where

$$\bar{Z}_{\mathcal{A}} = \bar{Z}_{\mathcal{A}}(\beta) \equiv \int_0^E dE_{\mathcal{A}}\,\bar{\omega}_{\mathcal{A}}(E_{\mathcal{A}})\exp(-\beta E_{\mathcal{A}}) \qquad (6.71)$$

and the upper integration limit can again be extended to ∞ with negligible error. Equation (6.70) now shows more explicitly that $\bar{p}_{\mathcal{A}}(E_{\mathcal{A}}|E)$ is a product of rapidly increasing and rapidly decreasing functions of $E_{\mathcal{A}}$. It is consequently sharply peaked about its maximum value $\bar{p}_{\mathcal{A}}(E_{\mathcal{A}}^*|E)$, as already discussed in Sects. 6.2 and 6.6.1, but it is instructive to see how this behavior emerges from the exponential factor in Eq. (6.70). To this end, we observe that $\bar{E}_{\mathcal{A}} \equiv \int dE_{\mathcal{A}}\,\bar{p}_{\mathcal{A}}(E_{\mathcal{A}}|E)E_{\mathcal{A}}$ and $\delta E_{\mathcal{A}}^2 \equiv \int dE_{\mathcal{A}}\,\bar{p}_{\mathcal{A}}(E_{\mathcal{A}}|E)(E_{\mathcal{A}} - \bar{E}_{\mathcal{A}})^2$ are simply related to the first and second derivatives of $\bar{Z}_{\mathcal{A}}$ via the relations

$$\bar{E}_{\mathcal{A}} = -\frac{\partial \log \bar{Z}_{\mathcal{A}}}{\partial \beta} \qquad (6.72)$$

$$\delta E_{\mathcal{A}}^2 = \frac{\partial^2 \log \bar{Z}_{\mathcal{A}}}{\partial \beta^2} = -\frac{\partial \bar{E}_{\mathcal{A}}}{\partial \beta} = \kappa T^2 C_V^{\mathcal{A}} \qquad (6.73)$$

Equation (6.73) agrees with Eq. (6.65) in the limit when $C_V^{\mathcal{B}} \gg C_V^{\mathcal{A}}$. Since $\bar{E}_{\mathcal{A}} \sim C_V^{\mathcal{A}} T \sim N_{\mathcal{A}}\kappa T$, Eq. (6.73) implies $\delta E_{\mathcal{A}}/\bar{E}_{\mathcal{A}} \sim 1/\sqrt{N_{\mathcal{A}}}$, so fluctuations in E_A are of order $1/\sqrt{N_A}$ and are therefore negligible in macroscopic systems.

Since $\bar{p}_A(E_A|E)$ is sharply peaked about the value $E_A = E_A^* \cong \bar{E}_A$, the only values of E_A which significantly contribute to the integral in Eq. (6.71) are those for which $|E_A - E_A^*| \lesssim \delta E_A$. The approximation

$$\bar{Z}_A \cong \delta E_A \, \bar{\omega}_A(E_A^*) \exp(-\beta E_A^*) \tag{6.74}$$

is therefore accurate to within a factor of order unity, which is logarithmically insignificant since Eq. (6.72) shows that $\log \bar{Z}_A$ is of order N. Thus

$$\log \bar{Z}_A = S_A(E_A^*) - \beta E_A^* + \log(\delta E_A/\Delta E) \tag{6.75}$$

where use has been made of Eq. (6.15). Moreover,

$$\log(\delta E_A/\Delta E) = \log(\delta E_A/\bar{E}_A) + \log(\bar{E}_A/\Delta E)$$

in which the first term is of order $\log N \ll N$ and the second is negligible as usual. Equation (6.75) therefore reduces to the simple convenient relation

$$\bar{S}_A = \beta \bar{E}_A + \log \bar{Z}_A = \log \bar{Z}_A - \beta \frac{\partial \log \bar{Z}_A}{\partial \beta} \tag{6.76}$$

where $\bar{S}_A \equiv S_A(E_A^*)$. Equation (6.76) shows that the entropy of system $\mathcal{A}$ in thermal equilibrium is simply related to $\log \bar{Z}_A$, and that the Helmholtz free energy in equilibrium is simply $\bar{F}_A \equiv F_A(E_A^*) = -\kappa T \log \bar{Z}_A$.

An alternative expression for $\bar{S}_A$ can be derived by taking the logarithm of Eq. (6.70), multiplying the result by $\bar{p}_A(E_A|E)$, and integrating over E_A. We thereby obtain

$$\beta \bar{E}_A + \log \bar{Z}_A = - \int dE_A \, \bar{p}_A(E_A|E) \log \frac{\bar{p}_A(E_A|E)}{\bar{\omega}_A(E_A)} \tag{6.77}$$

which combines with Eq. (6.76) to yield

$$\bar{S}_A = - \int dE_A \, \bar{p}_A(E_A|E) \log \frac{\bar{p}_A(E_A|E)}{\bar{\omega}_A(E_A)} \tag{6.78}$$

Equation (6.78) has precisely the form of Eq. (2.19) for the BGS entropy of a continuous probability density, and Eq. (6.70) is moreover a special case of the generalized canonical probability density of Eq. (5.31), which was obtained by maximizing the BGS entropy consistent with the mean value of the continuous state variable. In this chapter, however, we have deliberately refrained from invoking either the BGS entropy itself or the PME. The present development has been based entirely on the BP entropy and the EAPP hypothesis, and the fact that it has independently led to the BGS entropy and the probability density which maximizes it confirms and reinforces both the mathematical form of the BGS entropy and the PME. However, it has done so only within the specific context of macroscopic thermodynamic systems, whereas the previous considerations upon which the BGS entropy and the PME were based are of much greater generality and wider applicability. That generality was the rationale for discussing those concepts prior to the treatment of thermodynamic systems.

Chapter 7

THE CANONICAL AND GRAND CANONICAL PROBABILITY DISTRIBUTIONS

The terms *microcanonical*, *canonical*, and *grand canonical* were coined by Gibbs [17] to describe systems which are respectively either isolated, in thermal equilibrium, or in thermal and diffusional equilibrium with a reservoir. Gibbs applied those adjectives to the term *ensemble*, which denotes an imaginary collection of a very large number of mental replicas of a particular system of interest, in which the number of fictitious systems that occupy each state is proportional to its probability. The concept of an ensemble thenceforth became a conventional and well established feature of statistical mechanics, which however is unfortunate in some respects. Gibbs' benevolent intention was merely to facilitate the visualization of probability distributions by those unfamiliar with statistical concepts [20, 37]. Nowadays, however, students are normally introduced to basic probability theory well before they encounter statistical mechanics, and the concept of an ensemble has become a historical anachronism which is more likely to create confusion than to confer enlightenment. It seems preferable to focus attention directly on the system of interest and the uncertainty as to which of its states it occupies [37]. To this end and in that spirit, we eschew the term "ensemble" and use the terms microcanonical, canonical, and grand canonical exclusively to refer to the probability densities and distributions that describe the particular system of interest.

The probability densities considered in Chapter 6 describe statistical variations in the thermodynamic variables $E_{\mathcal{A}}$, $V_{\mathcal{A}}$, and/or $N_{\mathcal{A}}$ of system $\mathcal{A}$, but not the corresponding probability distributions $p_i^{\mathcal{A}}$ of its individual microstates i. In this chapter we extend the development of Chapter 6 to determine the form of the canonical and grand canonical microstate probability distributions $p_i^{\mathcal{A}}$ when the system of interest $\mathcal{A}$ is in equilibrium with a much larger system $\mathcal{B}$ which serves as a reservoir of heat and/or particles. In contrast to Chapter 6, however, we now no longer require system $\mathcal{A}$ itself to be large or macroscopic, although it very well may be. The resulting probability distributions therefore remain applicable when system $\mathcal{A}$ is small, indeed even when it consists of a mere single particle.

7.1 The Canonical Probability Distribution

We therefore proceed to reconsider the situation analyzed in Sect. 6.6.2 in which a system $\mathcal{A}$ is in thermal equilibrium with a much larger system $\mathcal{B}$, but we no longer require system $\mathcal{A}$ to be macroscopic in size. It remains understood that the coupling between the two systems is sufficiently weak that it does not alter their respective microstate energies. When system $\mathcal{A}$ is in microstate i its energy has the definite value $E_{\mathcal{A}} = E_i^{\mathcal{A}}$ and the energy of system $\mathcal{B}$ is $E_{\mathcal{B}} = E - E_i^{\mathcal{A}}$, where E is the constant total energy of the two systems together. The number of microstates accessible to system $\mathcal{B}$ is then

$$W_{\mathcal{B}}(E_{\mathcal{B}}) = W_{\mathcal{B}}(E - E_i^{\mathcal{A}}) = \exp S_{\mathcal{B}}(E - E_i^{\mathcal{A}}) \qquad (7.1)$$

Just as in Sect. 6.6.2, we may presume that $E_i^{\mathcal{A}} \ll E$, whereupon $S_{\mathcal{B}}(E - E_i^{\mathcal{A}})$ can be accurately approximated by

$$S_{\mathcal{B}}(E - E_i^{\mathcal{A}}) = S_{\mathcal{B}}(E) - \beta E_i^{\mathcal{A}} \qquad (7.2)$$

where $\beta = \partial S_{\mathcal{B}}(E)/\partial E = 1/(\kappa T)$, and T is the temperature of system $\mathcal{B}$ in the limit $E_i^{\mathcal{A}} \ll E$, in which $E_{\mathcal{B}} \cong E$. Combining

Eqs. (7.1) and (7.2), we obtain

$$W_{\mathcal{B}}(E - E_i^{\mathcal{A}}) = W_{\mathcal{B}}(E) \exp(-\beta E_i^{\mathcal{A}}) \tag{7.3}$$

The probability $p_i^{\mathcal{A}}$ that system $\mathcal{A}$ occupies microstate i is simply proportional to the number of microstates j of system $\mathcal{B}$ consistent with the energy $E_{\mathcal{B}} = E - E_i^{\mathcal{A}}$, which is just $W_{\mathcal{B}}(E - E_i^{\mathcal{A}})$ as given by Eq. (7.3). The probabilities $p_i^{\mathcal{A}}$ must of course sum to unity, so

$$p_i^{\mathcal{A}} = (1/Z_{\mathcal{A}}) \exp(-\beta E_i^{\mathcal{A}}) \tag{7.4}$$

where

$$Z_{\mathcal{A}} = Z_{\mathcal{A}}(\beta) \equiv \sum_i \exp(-\beta E_i^{\mathcal{A}}) \tag{7.5}$$

The mean energy of system $\mathcal{A}$ is therefore given by

$$\langle E_i^{\mathcal{A}} \rangle = \sum_i p_i^{\mathcal{A}} E_i^{\mathcal{A}} = -\frac{\partial \log Z_{\mathcal{A}}}{\partial \beta} \tag{7.6}$$

Equations (7.4) and (7.5) are identical in form to the canonical probability distribution and partition function previously encountered in Eqs. (5.35) and (5.36). Note that these results, like those of Sect. 6.6.2, have emerged without assuming either the functional form of the BGS entropy or the PME upon which Eqs. (5.35) and (5.36) were based. The present development thereby further confirms and supports both the BGS entropy and the PME, even for systems $\mathcal{A}$ which are not necessarily macroscopic in size.

It is instructive to confirm that these results are consistent with, and indeed essentially equivalent to, those of Sect. 6.6.2 when system $\mathcal{A}$ is macroscopic. To this end, we note that the average of an arbitrary function $f(E_i^{\mathcal{A}})$ can be written as

$$\langle f(E_i^{\mathcal{A}}) \rangle = \sum_i p_i^{\mathcal{A}} f(E_i^{\mathcal{A}}) = \int dE_{\mathcal{A}}\, p_{\mathcal{A}}(E_{\mathcal{A}}|E) f(E_{\mathcal{A}}) \tag{7.7}$$

where

$$p_{\mathcal{A}}(E_{\mathcal{A}}|E) \equiv \sum_i p_i^{\mathcal{A}}\, \delta(E_{\mathcal{A}} - E_i^{\mathcal{A}}) \tag{7.8}$$

is the true discontinuous probability density of E_A, which depends on E only through $\beta(E)$. Combining Eqs. (7.4) and (7.8), we obtain

$$p_A(E_A|E) = (1/Z_A)\,\omega_A(E_A)\exp(-\beta E_A) \tag{7.9}$$

where

$$\omega_A(E_A) \equiv \sum_i \delta(E_A - E_i^A) \tag{7.10}$$

is the true discontinuous density of states of system $\mathcal{A}$ per unit energy, in terms of which Eq. (7.5) can be rewritten as

$$Z_A(\beta) = \int dE_A\,\omega_A(E_A)\exp(-\beta E_A) \tag{7.11}$$

A comparison of Eqs. (7.9) and (7.11) with (6.70) and (6.71) shows that replacing $\omega_A(E_A)$ by its smooth continuous surrogate $\bar{\omega}_A(E_A)$ transforms $Z_A(\beta)$ into $\bar{Z}_A(\beta)$ and $p_A(E_A|E)$ into $\bar{p}_A(E_A|E)$. This replacement is justified when system $\mathcal{A}$ is macroscopic, in which case the discrete energies E_i^A are very closely spaced in comparison to $E_A \sim N_A/\beta$, whereupon $Z_A(\beta)$ becomes virtually identical to $\bar{Z}_A(\beta)$ and $\bar{p}_A(E_A|E)$ becomes operationally equivalent to $p_A(E_A|E)$. Conversely, however, if system $\mathcal{A}$ is not large enough for its microstate energies E_i^A to be closely spaced and densely distributed then $\omega_A(E_A)$ cannot be replaced by $\bar{\omega}_A(E_A)$, in which case Eqs. (7.4) and (7.5), or (7.9) and (7.11), must simply be used as they stand.

Similar considerations apply to the entropy S_A of system $\mathcal{A}$. Since the probabilities p_i^A are unequal, S_A must be evaluated as the BGS entropy of Eq. (2.10):

$$S_A = -\sum_i p_i^A \log p_i^A \tag{7.12}$$

Combining Eqs. (7.4), (7.6), and (7.12), we obtain

$$S_A = \beta\langle E_i^A\rangle + \log Z_A = \log Z_A - \beta\,\frac{\partial \log Z_A}{\partial \beta} \tag{7.13}$$

Comparison with Eq. (6.76) shows that the replacement $\omega_{\mathcal{A}}(E_{\mathcal{A}}) \to \bar{\omega}_{\mathcal{A}}(E_{\mathcal{A}})$ which transforms $Z(\beta)$ into $\bar{Z}(\beta)$ likewise transforms $S_{\mathcal{A}}$ into $\bar{S}_{\mathcal{A}}$. It is also interesting to observe in hindsight that the present results could actually have been inferred from those of Sect. 6.6.2 via the reverse replacement $\bar{\omega}_{\mathcal{A}}(E_{\mathcal{A}}) \to \omega_{\mathcal{A}}(E_{\mathcal{A}})$, whereby Eq. (6.70) is transformed into Eq. (7.9), which combines with Eq. (7.10) to become Eq. (7.8) with $p_i^{\mathcal{A}}$ given by Eq. (7.4). Similarly, Eq. (6.76) thereby transforms into Eq. (7.13) and thence into Eq. (7.12). However, this procedure would neither have demonstrated nor suggested that the present results remain valid even when system $\mathcal{A}$ is not macroscopic. Indeed, we reemphasize that the results of this section remain valid even when system $\mathcal{A}$ consists of a single particle, but only if its interaction energy with other particles is negligible as the derivation presumes. The latter condition is tantamount to a restriction to a single particle in an ideal gas.

A skeptical reader might wonder whether the exponential form of Eq. (7.4) has been prejudicially preordained by the superficially arbitrary decision to approximate $W_{\mathcal{B}}(E - E_i^{\mathcal{A}})$ in terms of a truncated Taylor series expansion of $S_{\mathcal{B}}(E - E_i^{\mathcal{A}}) = \log W_{\mathcal{B}}(E - E_i^{\mathcal{A}})$ rather than some other slowly varying function of $W_{\mathcal{B}}$. This concern is unfounded, however, as indicated by the observation that $(E - E_i)^N = E^N(1 - E_i/E)^N$ unambiguously tends to $E^N \exp(-NE_i/E)$ for large N when $E_i \ll E$. Moreover, the accuracy of Eq. (7.4) can be verified *a posteriori* by retaining the previously neglected quadratic term in Eq. (7.2), which is easily found to be $-\beta(E_i^{\mathcal{A}})^2/(2C_V^{\mathcal{B}}T)$. The net effect of this term is to multiply the exponent in Eq. (7.3) by a correction factor $1 + E_i^{\mathcal{A}}/(2C_V^{\mathcal{B}}T) \sim 1 + E_i^{\mathcal{A}}/(2E)$, which differs negligibly from unity since $E_i^{\mathcal{A}} \ll E$.

Different textbooks present various other derivations of the canonical probability distribution, some of which are unfortunately specious; *caveat lector!* An especially insidious subclass of such derivations is that based on the assumption that two weakly interacting systems

$\mathcal{A}$ and $\mathcal{B}$ in equilibrium with the same heat bath $\mathcal{R}$ are statistically independent, which is often asserted as self-evident. The canonical distribution then follows trivially after just a few lines of algebra. However, this essentially begs the question; statistical independence is a nontrivial assumption which requires justification. It is instructive to pursue such a justification along the same lines as the preceding derivation. The joint probability distribution $p_{ij}^{\mathcal{AB}}$ is now proportional to $W_{ij}^{\mathcal{R}} \equiv W_{\mathcal{R}}(E - E_i^{\mathcal{A}} - E_j^{\mathcal{B}})$, which clearly does not factor in general. Indeed, it evidently factors only when the Taylor series expansion of $\log W_{ij}^{\mathcal{R}}$ is truncated at the linear term, which is again justified only when the heat bath $\mathcal{R}$ is much larger than either system $\mathcal{A}$ or $\mathcal{B}$. One then obtains $W_{ij}^{\mathcal{R}} \cong W_{\mathcal{R}}(E)\exp(-\beta E_i^{\mathcal{A}})\exp(-\beta E_j^{\mathcal{B}})$ and thereby $p_{ij}^{\mathcal{AB}} = p_i^{\mathcal{A}} p_j^{\mathcal{B}}$. However, when the expansion is carried to second order $W_{ij}^{\mathcal{R}}$ contains the factor $\exp\{-\beta E_i^{\mathcal{A}} E_j^{\mathcal{B}}/(C_V^{\mathcal{R}} T)\}$, which cannot be expressed as a product of factors involving only $E_i^{\mathcal{A}}$ or $E_j^{\mathcal{B}}$. The assumption of statistical independence is therefore valid only when the heat bath is very large, and even then it is not immediately obvious *a priori*.

7.2 The Grand Canonical Probability Distribution

The preceding development can readily be generalized to the case in which system $\mathcal{A}$ can exchange particles as well as energy with system $\mathcal{B}$. The fixed total energy and number of particles of both systems together are then $E = E_{\mathcal{A}} + E_{\mathcal{B}}$ and $N = N_{\mathcal{A}} + N_{\mathcal{B}}$. As discussed in Sect. 5.3, the microstates of system $\mathcal{A}$ must now be labeled by the composite index $(N_{\mathcal{A}}, i)$, which is more conveniently displayed as an argument rather than a subscript. When system $\mathcal{A}$ is in microstate $(N_{\mathcal{A}}, i)$, it contains precisely $N_{\mathcal{A}}$ particles and its energy has the definite value $E_{\mathcal{A}} = E_{\mathcal{A}}(N_{\mathcal{A}}, i)$, so system $\mathcal{B}$ has energy $E_{\mathcal{B}} = E - E_{\mathcal{A}}$ and contains $N_{\mathcal{B}} = N - N_{\mathcal{A}}$ particles. The number of microstates $(N_{\mathcal{B}}, j)$ accessible to system $\mathcal{B}$ is then

$$W_{\mathcal{B}}(E - E_{\mathcal{A}}, N - N_{\mathcal{A}}) = \exp S_{\mathcal{B}}(E - E_{\mathcal{A}}, N - N_{\mathcal{A}}) \qquad (7.14)$$

Since system $\mathcal{A}$ is much smaller than system $\mathcal{B}$, we may presume as before that $E_{\mathcal{A}} \ll E$ and $N_{\mathcal{A}} \ll N$, so that $S_{\mathcal{B}}(E - E_{\mathcal{A}}, N - N_{\mathcal{A}})$ can be accurately approximated by

$$S_{\mathcal{B}}(E - E_{\mathcal{A}}, N - N_{\mathcal{A}}) = S_{\mathcal{B}}(E, N) - \beta E_{\mathcal{A}} - \eta N_{\mathcal{A}} \qquad (7.15)$$

where $\beta = \partial S_{\mathcal{B}}(E, N)/\partial E = 1/(\kappa T)$, $\eta = \partial S_{\mathcal{B}}(E, N)/\partial N = -\beta\mu$, and T and μ are respectively the temperature and chemical potential of the reservoir $\mathcal{B}$ in the limit $(E_{\mathcal{A}}, N_{\mathcal{A}}) \ll (E, N)$, in which $(E_{\mathcal{B}}, N_{\mathcal{B}}) \cong (E, N)$. Combining Eqs. (7.14) and (7.15), we obtain

$$W_{\mathcal{B}}(E - E_{\mathcal{A}}, N - N_{\mathcal{A}}) = W_{\mathcal{B}}(E, N) \exp(-\beta E_{\mathcal{A}} - \eta N_{\mathcal{A}}) \qquad (7.16)$$

The probability $p_{\mathcal{A}}(N_{\mathcal{A}}, i)$ that system $\mathcal{A}$ contains $N_{\mathcal{A}}$ particles and is in microstate i is proportional to $W_{\mathcal{B}}(E - E_{\mathcal{A}}(N_{\mathcal{A}}, i), N - N_{\mathcal{A}})$, and the coefficient of proportionality is determined by normalization as usual. It follows that

$$\begin{aligned} p_{\mathcal{A}}(N_{\mathcal{A}}, i) &= (1/\Xi) \exp\{-\beta E_{\mathcal{A}}(N_{\mathcal{A}}, i) - \eta N_{\mathcal{A}}\} \\ &= (1/\Xi)\, \lambda^{N_{\mathcal{A}}} \exp\{-\beta E_{\mathcal{A}}(N_{\mathcal{A}}, i)\} \end{aligned} \qquad (7.17)$$

where $\lambda \equiv \exp(\beta\mu)$ and

$$\Xi = \Xi(\beta, \eta) \equiv \sum_{N_{\mathcal{A}}, i} \exp\{-\beta E_{\mathcal{A}}(N_{\mathcal{A}}, i) - \eta N_{\mathcal{A}}\} \qquad (7.18)$$

From this point on the only vestigal features or remnants of the reservoir $\mathcal{B}$ that remain in the equations are the parameters β and η. Thus the subscript $\mathcal{A}$ can henceforth be suppressed with the understanding that system $\mathcal{A}$ is the system of interest, and the cumbersome argument $(N_{\mathcal{A}}, i)$ representing the microstate index can be converted into a subscript Nk. When this is done, Eqs. (7.17) and (7.18) become identical to the grand canonical probability distribution and partition function previously encountered in Eqs. (5.42) and (5.43), which again immediately imply Eqs. (5.44) and (5.45). The probabilities

p_{Nk} are unequal, so the entropy of the system is again given by the BGS entropy of Eq. (2.10):

$$S = -\sum_{Nk} p_{Nk} \log p_{Nk} \tag{7.19}$$

It is straightforward to verify that Eqs. (5.42)–(5.45) and (7.19) imply Eqs. (5.46)–(5.48), so that Eqs. (5.42)–(5.48) in their entirety have now been independently confirmed and derived by a completely different method in which the PME was not assumed or invoked. The only minor difference is that the roles of $(\bar{E}, \bar{N})$ and (β, η) as independent and dependent variables have been interchanged; in Chapter 5 the parameters (β, η) were determined as functions of $(\bar{E}, \bar{N})$ by Eqs. (5.47) and (5.48), whereas in the present context (β, η) are properties of the reservoir $\mathcal{B}$ which then determine $(\bar{E}, \bar{N})$ via Eqs. (5.44) and (5.45). Of course, the functional relationships between (β, η) and $(\bar{E}, \bar{N})$ remain the same either way.

7.3 Interlude

It is remarkable that we have been able to progress this far without actually defining or enumerating the microstates of thermodynamic systems. The development up to this point merely depends upon the existence of the microstates k, and is independent of their characteristics or structural details. The preceding results and formulae are consequently possessed of great generality; they apply to systems composed of either independent or interacting particles in both classical and quantum mechanics, and in the latter case to both bosons and fermions. However, a detailed consideration of the microstates cannot be postponed indefinitely; they must be precisely identified, labeled, and characterized before quantitative results can be extracted from the formalism. This is a nontrivial and rather intricate task for the many-particle systems of present interest, and it consequently requires a detailed discussion and development which will occupy the final two chapters of this book. This development is not unduly difficult, but it requires concentration and a clear understanding of the concepts which underlie the mathematical manipulations. Those concepts will accordingly be discussed in some detail.

Readers with some previous exposure to statistical mechanics may be surprised that up to this point we have had no occasion to make use of combinatorial concepts and arguments involving permutations, combinations, binomial and multinomial coefficients and distributions, etc. In contrast, such concepts are typically introduced at the outset and play a prominent role in introductory treatments. The reason for this disparity is that appearances to the contrary notwithstanding, combinatorial concepts are not in fact essential ingredients in the general formalism of entropy, which indeed is actually independent of them as the preceding development attests. Combinatorial concepts do not arise until the general formalism is specialized to composite systems, such as many-particle systems, which are made up of simpler systems which share microstates. Even then, such concepts are in essence merely bookkeeping devices used to classify and enumerate the microstates. Of course, many-particle systems arguably constitute the single most important application of entropy, and it is essential to properly characterize and enumerate their microstates. However, this should not be allowed to obscure or eclipse a broader perspective of entropy as a statistical construct of more general significance and applicability. The organization of this book has been designed to accurately reflect the hierarchial logical structure of the material, thereby making it easier to discern which concepts are required for which purposes. Conversely, conventional treatments which begin with combinatorial considerations are liable to create an erroneous impression of their role and significance.

Chapter 8

MANY-PARTICLE SYSTEMS

Many of the most interesting and important applications of entropy are to composite systems composed of simpler subsystems which either are, or can be thought of as, various types of particles. It is therefore convenient to refer to them as such, with the understanding that in some cases it may be appropriate to interpret the term "particle" in a generalized sense. The number of particles N in such systems is typically very large, so they are commonly referred to as "many-particle" systems, but it is well to keep in mind that the concept of entropy (and consequently much of the general formalism developed prior to Chapter 6) is equally applicable to both simple and composite systems of any type or number of subsystems, including particulate systems of small N. In addition to being of intrinsic interest in their own right, systems of small N provide the clearest illustration of the additional concepts required to apply the general formalism of entropy to systems of arbitrary N, including macroscopic thermodynamic systems containing $N \sim 10^{23}$ particles.

As stated in the Introduction, we restrict attention to pure systems composed of particles which are all of the same type or species. Since the particles are identical, they all possess the same set of single-particle states (i.e., states accessible to a single isolated particle). If the N particles in the system do not interact with one another then each particle simply occupies one of those single-particle states, and conversely each such state is either unoccupied or occupied by one or more particles. Thus the particles can be regarded as sharing the single-particle states, and the microstates of the N-particle system as a whole can be defined in terms of the single-particle states occupied by the individual particles. This description is valid for systems of non-interacting particles (i.e., ideal gases) in both classical and quantum mechanics. Moreover, it remains valid in classical mechanics even for systems of interacting particles, in which each particle continues to occupy the definite single-particle state $(\mathbf{q}, \mathbf{p})$ defined by its own coordinates $\mathbf{q}$ and momentum $\mathbf{p}$. Of course, interparticle interactions induce correlations between the particles and the states they occupy, so interacting particles are not in general statistically independent, but this does not in any way invalidate the description of their states.

Unfortunately, the simple physical picture in which each particle occupies a definite single-particle state is no longer strictly valid for interacting particles in quantum mechanics. However, it nevertheless remains very useful and plays a prominent role in that context as both an approximation and a metaphor, as exemplified by atomic and molecular orbital theory. Moreover, the critical conceptual issues related to the statistical implications of particle distinguishability and indistinguishability are most clearly addressed, understood, and resolved in terms of the occupancy of single-particle states. The present treatment of many-particle systems is accordingly based on the premise that each particle in the system occupies one of its accessible single-particle states, with the understanding that this simple physical picture becomes approximate and requires appropriate modification for systems of interacting particles in quantum mechanics. Those modifications will be indicated in Sect. 8.8. From a general point of view, however, these issues are technical details which should not be allowed to obscure the fact that the general relations derived in previous chapters remain valid as they stand for systems of either independent or interacting particles in both classical and quantum mechanics. Those relations determine the probabilities of the multi-particle microstates of the entire N-particle system, regardless of how those states are labeled and whether they are or are not simply related to the single-particle states. The entropy of the many-particle system is then given by the BP entropy if its multi-particle microstates are equally probable, or the BGS entropy if they are not. In the former case, it is simply necessary to count the microstates, which however is not always easy inasmuch as they are typically subject to various constraints. In any case, the BGS entropy is always valid, since it automatically reduces to the BP entropy for equally probable microstates.

It is by no means trivial or obvious *a priori* how to properly define, describe, label, and/or enumerate the microstates of a system of N particles in terms of the shared single-particle microstates. To accomplish this task, it is necessary to identify and address certain essential conceptual issues which are peculiar to composite systems and thus did not arise in the preceding chapters. In brief, those issues involve the related concepts of (a) distinguishable vs. indistinguishable particles; (b) generic vs. specific multi-particle states; and (c) the interesting and perhaps counterintuitive result that the EAPP hypothesis and the PME imply that indistinguishable particles are not in general statistically independent, even when they do not directly interact with one another (i.e., when their interaction energy is negligible). Before attempting to discuss these concepts in general, it is instructive to illustrate them in a simple context where their essential features are more clearly revealed and easily understood. Fortunately, these important concepts already manifest themselves in the simplest nontrivial multi-particle system, namely a system of two particles, each of which possesses two single-particle states. Such a system is epitomized by two coins. A careful analysis of the states and statistical properties of a system of two coins is remarkably enlightening, and leads to several more general insights. Such an analysis is therefore given in some detail in the next section.

8.1 A Tale of Two Coins

8.1.1 Distinguishable Coins

We temporarily rescind our restriction to identical particles in order to examine the statistical properties of two coins of distinct types, for example a nickel and a dime. Each coin has two single-particle states a, namely heads (a = H) and tails (a = T). Both coins are presumed to be fair (unbiased), so the probabilities of their respective single-particle states a are the same for both coins and are given by $p(\mathrm{H}) = p(\mathrm{T}) = 1/2$. The states of the nickel-dime system are then the ordered pairs (a, b) $(a, b = \mathrm{H}, \mathrm{T})$, where a is the state of the nickel and b is the state of the dime. Thus the nickel-dime system has four states, namely (H, H), (T, T), (H, T), and (T, H). In the terminology of Gibbs [17], the states (a, b) are referred to as *specific states*, because they specify precisely which state is occupied by each of the coins. The probability $p(a, b)$ of the specific state (a, b) is the joint probability that the nickel is in state a and the dime is simultaneously in state b, which allows for the possibility that the states of the two coins are correlated; i.e., that the state occupied by one coin influences the state of the other. However, if the coins are flipped individually and are not allowed to come into contact, they can be presumed to be uncorrelated (i.e., statistically independent), in which case the joint probability factors: $p(a, b) = p(a)\,p(b)$. The resulting value of $p(a, b)$ is then $(1/2)\times(1/2) = 1/4$ independently of a and b, so the specific states (a, b) are equally probable. The entropy of the nickel-dime system is then simply the BP entropy $S = \log 4$, which of course is also the maximum possible BGS entropy of a system with four states. Thus the EAPP hypothesis and the PME are valid for this system.

8.1.2 Indistinguishable Coins

Now consider two indistinguishable coins, for example two uncirculated dimes minted at the same time in the same production run.

Of course, dimes are macroscopic objects which are obviously not absolutely indistinguishable on a microscopic or molecular level. However, our advanced technological society has the capability to manufacture them to such high precision that they could only be distinguished by equally high precision instrumentation, and hence are effectively indistinguishable for all practical purposes outside a research laboratory environment. We shall suppose that this has been done and that we have been assigned the task of characterizing the statistical properties of two such coins as best we can without access to sophisticated instrumentation. (It is also understood that we are not allowed to mark or label the coins in any way, for if this were done they would no longer be indistinguishable and would become logically equivalent to the nickel-dime system analyzed above.) We again presume that both coins are fair and unbiased, so their single-particle states H and T are again equally probable; i.e., $p(\mathrm{H}) = p(\mathrm{T}) = 1/2$.

We now proceed to consider the two coins together as a two-particle system. What now defines the states of this system? Since we no longer have any way to distinguish between the coins (i.e., to tell which coin is which), it is no longer meaningful to attempt to specify which coin is in which single-particle state. The specific states (a, b) therefore no longer exist, because they require one to specify which particular coins occupy states a and b. We therefore revert to the basic definition of the state of a system as a complete specification of its configuration at the level of the description employed, which in the present context is defined by the variables H and T. The only configurations of the two indistinguishable coins which our senses are capable of perceiving and distinguishing between are whether they are both heads, both tails, or one of each. Those three possibilities are mutually exclusive and exhaustive, and can be respectively represented by the unordered *multisets* $\{\mathrm{H}, \mathrm{H}\}$, $\{\mathrm{T}, \mathrm{T}\}$, and $\{\mathrm{H}, \mathrm{T}\} = \{\mathrm{T}, \mathrm{H}\}$. (A multiset differs from a set in that its elements can be repeated. The multiplicity of each element is the number of times it appears in the multiset.) The multisets $\{a, b\}$

therefore define the three possible states of the two-particle system. These states are referred to as *generic states* because they specify which single-particle states the coins occupy, but not which coin is in which state. We reemphasize that they constitute the true microstates of a system of two indistinguishable coins, because they provide a full and complete specification of its configuration.

Even though the two coins are indistinguishable and we cannot tell them apart, they still retain their individual existence as separate physical objects, so we can still flip them individually (e.g., sequentially) in such a way that they do not come into contact with or otherwise influence each other. When this is done, the two coins can again be presumed to be statistically independent. The only way the generic state $\{\mathrm{H}, \mathrm{H}\}$ can occur is if both coins come up heads, so its probability is $p\{\mathrm{H}, \mathrm{H}\} = p(H)\, p(H) = 1/4$. Similarly, $p\{\mathrm{T}, \mathrm{T}\} = p(T)\, p(T) = 1/4$. The probabilities of the three generic states must sum to unity, so the probability of the generic state $\{\mathrm{H}, \mathrm{T}\}$ is $p\{\mathrm{H}, \mathrm{T}\} = 1 - p\{\mathrm{H}, \mathrm{H}\} - p\{\mathrm{T}, \mathrm{T}\} = 1/2$. Thus we see that in contrast to the specific states of a system of two distinguishable coins, the probabilities of the generic states of a system of two indistinguishable coins are unequal when the coins are statistically independent. The entropy of the system must therefore be evaluated as its BGS entropy, which is given by

$$S = -\sum_{\{a,b\}} p\{a, b\} \log p\{a, b\} = \tfrac{1}{2} \log 8 = \log \sqrt{8} \qquad (8.1)$$

This is smaller than the maximum BGS entropy for a system with three states, which is simply $\log 3$. Thus naively applying the EAPP hypothesis or the PME to this system would produce incorrect results. The explanation for this apparent anomaly is that the statistical independence of the two coins is tantamount to an additional constraint on the system (or additional knowledge about the system, in terms of information theory). That additional constraint or knowledge invalidates the EAPP hypothesis, and since it uniquely determines the probabilities $p\{a, b\}$ the PME is no longer applicable. The entropy or uncertainty of the system is then smaller than its maximum possible value of $\log 3$ for a system with three states, which in turn is smaller yet than it would be if the particles were distinguishable and the number of states were four rather than three.

The preceding analysis conversely implies that *if the EAPP hypothesis* is *valid for the true generic states of a system of two indistinguishable fair coins, then the coins are not statistically independent*, for if they were the probabilities $p\{a, b\}$ would be unequal as shown above. Under what circumstances might such a situation be expected to occur? Statistical dependence generally results from some sort of interaction between the particles, albeit possibly of an obscure or unknown nature. Such interactions typically occur when the particles are in close proximity; e.g., if the two coins are simultaneously flipped together in such a way that they are allowed to interfere with each other. In situations where interparticle interactions seem likely to preclude statistical independence but no information is available about those interactions and their effects, it is natural to invoke the EAPP hypothesis or the PME and compare the resulting predictions with the observed behavior of the system. In the present context of two indistinguishable coins, this procedure implies equal probabilities $p\{a, b\} = 1/3$, an entropy of $S = \log 3$, and that the coins are statistically dependent. It may be of interest to note that this situation constitutes the simplest example of Bose-Einstein statistics, although it is too simple to exhibit the further interesting features thereof. The more familiar Bose-Einstein distribution in statistical thermodynamics will be derived in Sect. 8.8.

8.1.3 Discussion

The concepts illustrated by this deceptively simple system are more generally applicable to systems of an arbitrary number of particles N which share a finite or infinite set of discrete single-particle states. However, their application to such systems requires a correspondingly more general mathematical formulation, the development of which will be undertaken in the next section. But first it is well to recapitulate and reexpress in more general terms the essential features and implications of the preceding analysis, which are so fundamental and important that they can hardly be overemphasized:

(a) The essential difference between specific and generic multi-particle microstates is that the former specify precisely *which* particles occupy each single-particle microstate, whereas the latter specify *how many* particles occupy each single-particle microstate.

(b) The true multi-particle microstates of a system of distinguishable particles are its specific states.

(c) The true multi-particle microstates of a system of indistinguishable particles are its generic states.

(d) The entropy of a multi-particle system must always be computed in terms of the probabilities of its true microstates as defined by (b) and (c) above.

Unfortunately, the clear unambiguous simplicity of these basic points is sometimes obscured by the fact that it is also useful and customary to define generic and specific microstates in systems of distinguishable and indistinguishable particles, respectively, so that both types of systems are regarded as possessing both types of microstates. The generic states of distinguishable particles simply represent convenient groups of specific states. In contrast, specific states do not strictly exist in systems of indistinguishable particles, but they can nevertheless be defined as the specific states of a corresponding fictitious or imaginary system which differs from the real system only in that the particles are labeled. Specific states are easier to describe and work with than generic states, so they are useful for purposes of enumerating states and computing their probabilities. However, the use of fictitious specific states to describe indistinguishable particles creates an insidious danger of confusion and errors resulting from a failure to clearly draw or fully comprehend the distinction between the real system and its fictitious counterpart. In particular, it is nonsensical to evaluate the entropy of a system of indistinguishable particles in terms of the probabilities of its fictitious specific microstates, which is unfortunately a common error. In fact, it is this very error which gives rise to the famous Gibbs paradox and its associated infamous $1/N!$ correction factor, as will be discussed in due course.

8.2 Systems of Indistinguishable Particles

Having digested the enlightening insights gleaned from the Tale of Two Coins, we resume the main development by reimposing our original restriction to identical particles, which will remain in effect for the duration of the book. As stated in the Preface and Introduction, we regard all identical particles, both classical and quantum, as inherently indistinguishable for reasons to be further elaborated in Chapter 9. The following development is accordingly restricted to indistinguishable particles. Readers who remain unconvinced that identical classical particles are indistinguishable are hereby encouraged to reserve judgment on that question for the time being and reconsider it after reading Chapter 9.

Thus we proceed to explore the statistical properties of a system of N indistinguishable particles, including its specific and generic microstates, the relations between the two, their respective probabilities, and the entropy of the system as a whole. The first task at hand is to formulate suitable mathematical representations of the specific and generic multi-particle microstates of the system. As discussed in Section 8.1.3, this task is problematical, not to mention conceptually incongruous, due to the fact that specific microstates do not strictly exist in systems of indistinguishable particles. The reason is that such states refer to specific particles and therefore require the particles to be labeled, whereas indistinguishable particles carry no labels, and as a matter of principle they cannot *be* labeled, for if they were they would no longer be indistinguishable. This principle certainly applies to the particles in the actual system under consideration, but to borrow a felicitous phrase from Gibbs, it hardly applies to creations of the imagination. As discussed in Sect. 8.1.3, we are free to invent a hypothetical, fictitious, or imaginary system of N *distinguishable* particles which differs from the real system of interest only in that the N particles have been mentally furnished with passive labels $\alpha = 1, 2, \cdots, N$. By definition, we stipulate that (a) the labels α are the only distinguishing features of the fictitious distinguishable particles, which are otherwise identical; and (b) all of their other properties, including their single-particle microstates and the attributes (e.g., energy) and probabilities thereof, are identical to those of the real indistinguishable particles in the actual system of interest. The intrinsic or objective properties of the fictitious system are therefore likewise identical to those of the real system, and hence are invariant to permutations of the particle labels α, which can therefore be interchanged, reassigned, or redefined at will. This also follows from the fact that since the particles are otherwise identical, the labels assigned to them are entirely arbitrary, so all such assignments must be equivalent.

Since the real system of indistinguishable particles does not possess specific states of its own, we are further free to simply *define* its specific states as those of the corresponding fictitious system of distinguishable particles. This definition is logically unobjectionable but somewhat hazardous, because it requires one to remain cognizant at all times that the specific states thus defined are not objective attributes of the real system; they are rather mere mental constructs which must not be invested with more significance than they logically possess. They are really nothing more than a convenient mathematical device useful for counting states and computing probabilities, and it is essential not to misinterpret them as the true multi-particle microstates of the system. As discussed in Sect. 8.1, the true microstates are the generic states, and the entropy must accordingly be evaluated in terms of their probabilities. (Conversely, however, in a *real* system of labeled but otherwise identical particles the *specific* states would be the true microstates, and their probabilities would determine the entropy.)

8.3 Specific and Generic Microstates

The *specific* microstate occupied by an N-particle system is defined by specifying which particular particles α (if any) occupy each single-particle state k, or equivalently which state k_α is occupied by each particle α. The latter representation is simpler and more convenient, and is represented by the ordered N-tuple $\mathbf{k} \equiv (k_1, k_2, \cdots, k_N)$. The *generic* microstate occupied by an N-particle system is defined by specifying the number of particles n_k which occupy each single-particle state k. The quantities n_k are referred to as the occupation numbers or populations of the states k, and must satisfy the obvious constraint

$$\sum_k n_k = N \tag{8.2}$$

The occupation numbers are well defined even when the particles are not labeled, but in the present context they can be expressed as

$$n_k = \sum_{\alpha=1}^{N} \delta(k, k_\alpha) \tag{8.3}$$

where $\delta(i, j)$ is the Kronecker delta. If their values are obtained from Eq. (8.3) then Eq. (8.2) is automatically satisfied. It is convenient to collectively denote the occupation numbers n_k, and thereby the generic microstate itself, by $\mathbf{n} \equiv (n_1, n_2, \cdots, n_k, \cdots)$. The number of components of $\mathbf{n}$ is therefore infinite when the number of states k is infinite, but Eq. (8.2) implies that at most N of them are nonzero.

The generic microstate of the system can alternatively be represented by the multiset $\{\mathbf{k}\} \equiv \{k_1, k_2, \cdots, k_N\}$ of cardinality N, which is simply an unordered list of the occupied single-particle microstates, in which each state k in the list appears n_k times. Thus the elements of $\{\mathbf{k}\}$ are not necessarily distinct, but since there are N of them they can be arbitrarily labeled by the same integers $\alpha = 1, 2, \cdots, N$ used to label the particles in the fictitious system. In the multiset context, however, the subscripts α no longer refer to, represent, or identify specific particles. This is clearly illustrated by the multisets used to represent the generic states in the Tale of Two Coins. Thus, although the notation might suggest otherwise, the multiset $\{k_1, k_2, \cdots, k_N\}$ is independent of the labeling of the particles in the fictitious system, and indeed remains well defined even in the absence of the fictitious system; i.e., when the particles are not labeled at all. This simply reflects the fact that $\mathbf{n}$ and $\{\mathbf{k}\}$ are equivalent ways of representing the same information, and either of them determines and is determined by the other. For general use, the occupation numbers $\mathbf{n}$ are usually more convenient than the multisets $\{\mathbf{k}\}$, but not always.

The multiplicity $M(\mathbf{n})$ of the generic microstate $\mathbf{n}$ is defined as the number of specific states $\mathbf{k} = (k_1, k_2, \cdots, k_N)$ consistent with it. This is simply the number of ways in which N distinct particles can be assigned to the single-particle states so that state k contains n_k particles, or equivalently the number of distinct permutations of the N elements in the multiset $\{\mathbf{k}\}$. That number is just the familiar multinomial coefficient

$$M(\mathbf{n}) = \frac{N!}{\prod_k n_k!} = \frac{N!}{\mathbf{n}!} \tag{8.4}$$

where we have introduced the useful compact multi-index notation $\mathbf{n}! \equiv \prod_k n_k!$ (warning: some authors denote the multinomial coefficient itself by $\mathbf{n}!$). Note that $\mathbf{n}! = 1$ and $M(\mathbf{n}) = N!$ when none of the single-particle microstates k is occupied by more than one particle. Since the variable $\mathbf{n}$ explicitly specifies the number of particles in each state k, even unoccupied states for which $n_k = 0$, the product $\mathbf{n}! = \prod_k n_k!$ likewise extends over all states k, and is therefore an infinite product when the number of single-particle microstates is infinite. However, since $0! = 1$ this is equivalent to a finite product over occupied states only, which contains at most N factors. When the numbers of particles and single-particle states are both very large, it is clear that $M(\mathbf{n})$ is likewise ordinarily very large, in which case the specific states are much more numerous than the generic states. However, they both remain microstates which require a very large number of variables for their unique specification. Some authors confuse the generic microstates with macrostates, which is unfortunately a common egregious misconception, albeit less egregious than misinterpreting the specific states as the true microstates of the system.

8.4 Multi-Particle Probabilities

Since the fictitious particles are identical except for their labels, the probability $p_\alpha(k)$ that particle α occupies the single-particle microstate k is the same for all particles, so it depends only on k and is independent of α. It can therefore be denoted by $p(k)$, and clearly represents the probability that any one of the indistinguishable particles in the real system occupies state k. The probability $p_s(\mathbf{k})$ that the fictitious system occupies the specific microstate $\mathbf{k}$ is the joint probability that particles $1, 2, \cdots, N$ simultaneously occupy states $k_1, k_2, \cdots, k_N$ respectively, which allows for the possibility that the states of the particles are correlated. Since all intrinsic properties of the fictitious system are unaffected by interchanging the particle labels, $p_s(\mathbf{k})$ is invariant to permutations of its arguments and is therefore a symmetric function of $(k_1, k_2, \cdots, k_N)$. This implies that $p_s(\mathbf{k})$ actually depends only on the unordered multiset $\{\mathbf{k}\}$, which in turn depends only on the generic microstate $\mathbf{n}$, so it can equally well be written as either $p_s\{\mathbf{k}\}$ or $p_s(\mathbf{n})$.

The fact that $p_s(\mathbf{k})$ depends only on $\{\mathbf{k}\}$ or $\mathbf{n}$ implies that all the specific microstates $\mathbf{k}$ consistent with a particular generic microstate $\mathbf{n}$ have the same probability. The probability $p_g(\mathbf{n})$ that the system occupies the generic microstate $\mathbf{n}$ is therefore simply given by

$$p_g(\mathbf{n}) \;=\; M(\mathbf{n})\, p_s(\mathbf{k}) \;=\; \frac{N!}{\mathbf{n}!}\, p_s(\mathbf{k}) \tag{8.5}$$

Thus either $p_s(\mathbf{k})$ or $p_g(\mathbf{n})$ determines the other. Their actual values depend on the detailed properties of the system and its microstates, especially the energy $E(\mathbf{n})$, and the constraints imposed upon it (e.g., constant energy, thermal equilibrium, etc.). Those probabilities are to be determined by means of the relations derived in previous chapters. For this purpose, the rather imprecisely defined system states k in previous chapters must be identified with the generic states $\mathbf{n}$, since the latter constitute the true microstates of the system. Once the probabilities $p_g(\mathbf{n})$ have thereby been determined, the entropy of the system is simply given by its BGS entropy

$$S \;=\; -\sum_{|\mathbf{n}|=N} p_g(\mathbf{n}) \log p_g(\mathbf{n}) \tag{8.6}$$

where the summation extends over all values of the occupation numbers n_k that satisfy Eq. (8.2), which has been indicated by introducing the further multi-index notation $|\mathbf{n}| \equiv \sum_k n_k$, in terms of which Eq. (8.2) assumes the more compact form $|\mathbf{n}| = N$.

It is instructive to compare S with the entropy which the fictitious system of distinguishable particles would possess if it and they were real. Its true multi-particle microstates would then be the specific states $\mathbf{k}$, and its entropy would be given by

$$S_d \;=\; -\sum_{\mathbf{k}} p_s(\mathbf{k}) \log p_s(\mathbf{k}) \;=\; -\sum_{|\mathbf{n}|=N} M(\mathbf{n})\, p_s(\mathbf{k}) \log p_s(\mathbf{k}) \tag{8.7}$$

in which the subscript d stands for "distinguishable," and it must be remembered that $p_s(\mathbf{k})$ actually depends only upon $\mathbf{n}$. Combining Eqs. (8.5)–(8.7), we obtain

$$S_d \;=\; S + \sum_{|\mathbf{n}|=N} p_g(\mathbf{n}) \log M(\mathbf{n}) \tag{8.8}$$

If $p_g(\mathbf{n})$ is negligible whenever any single-particle state k is multiply occupied (i.e., $n_k \geq 2$), then the only values of $\mathbf{n}$ which significantly contribute to the sum in Eq. (8.8) are those in which n_k is either zero or unity for all k. One would intuitively expect this to be the case when $W_1(T) \gg N$, where $W_1(T)$ is the number of thermally accessible single-particle microstates; i.e, microstates k with energies $\varepsilon_k \lesssim \kappa T$. Since $0! = 1! = 1$, the multinomial coefficient $M(\mathbf{n})$ then reduces to $N!$ and Eq. (8.8) becomes

$$S_d = S + \log N! \tag{8.9}$$

This shows that S could have been computed by first computing S_d and then subtracting a "correction term" $\log N!$. This term is simply the logarithm of the infamous $1/N!$ correction factor used to resolve the Gibbs paradox. However, this correction term or factor is not arbitrary or *ad hoc*; it simply reflects the fact that the true multi-particle microstates of a system of indistinguishable particles are its generic states, not its fictitious specific states. If its entropy is properly computed using Eq. (8.6) in the first place, there is no paradox and nothing to correct. However, Eq. (8.9) shows that the incorrect use of S_d instead of S would nevertheless correctly predict the temperature and pressure, but not the chemical potential, since the term $\log N!$ drops out when S is differentiated with respect to energy or volume.

Equation (8.5) is useful for addressing questions related to statistical correlations between the particles. When the particles are statistically independent, their states are uncorrelated and $p_s(\mathbf{k})$ factors into the product of the single-particle probabilities $p(k_\alpha)$; i.e.,

$$p_s(\mathbf{k}) = \prod_{\alpha=1}^{N} p(k_\alpha) \tag{8.10}$$

Each factor of $p(k)$ appears n_k times in the above product, so Eq. (8.10) can be rewritten in the form

$$p_s(\mathbf{k}) = \prod_{k} p_k^{n_k} \tag{8.11}$$

where $p_k \equiv p(k)$. Combining Eqs. (8.5) and (8.11), we obtain

$$p_g(\mathbf{n}) = M(\mathbf{n}) \prod_k p_k^{n_k} = \frac{N!}{\mathbf{n}!} \prod_k p_k^{n_k} \tag{8.12}$$

which is just the well known multinomial distribution. The fact that $p_g(\mathbf{n})$ is properly normalized can be confirmed by summing Eq. (8.12) over $\mathbf{n}$ subject to the constraint of Eq. (8.2), and making use of the multinomial theorem and the fact that $\sum_k p_k = 1$. Equation (8.12) provides a test for statistical dependence: if $p_g(\mathbf{n})$ is not related to the single-particle probabilities p_k by Eq. (8.12) then the particles are not statistically independent. Conversely, if the particles are statistically independent and their single-particle probabilities p_k are known, Eq. (8.12) determines $p_g(\mathbf{n})$, which then determines the entropy S via Eq. (8.6) and thereby the thermodynamic properties of the system via Eqs. (6.29), (6.41), (6.51), and (6.54).

8.5 Ideal Gases: The Boltzmann Distribution

Since the generic states $\mathbf{n}$ are the true microstates of the system, its canonical partition function is simply

$$Z_N(\beta) = \sum_{|\mathbf{n}|=N} \exp\{-\beta E(\mathbf{n})\} = \sum_{\mathbf{k}} \frac{\exp\{-\beta E(\mathbf{k})\}}{M(\mathbf{n})} \tag{8.13}$$

where $E(\mathbf{n}) = E(\mathbf{k})$ is the energy of the N-particle system as a function of either $\mathbf{n}$ or $\mathbf{k}$, and the factor of $1/M(\mathbf{n})$ accounts for the fact that each term in the summation over $\mathbf{n}$ is repeated $M(\mathbf{n})$ times in the summation over $\mathbf{k}$. When the number density N/V of particles is sufficiently low and the temperature T is sufficiently high, the potential energy of interaction between the particles is negligible compared to their kinetic energy and can be neglected. The energy is then simply the sum of the energies of the individual particles:

$$E(\mathbf{n}) = \sum_k n_k \, \varepsilon_k = E(\mathbf{k}) = \sum_{\alpha=1}^{N} \varepsilon(k_\alpha) \tag{8.14}$$

where $\varepsilon_k = \varepsilon(k)$ denotes the energy of single-particle microstate k, and use has been made of Eq. (8.3). We further assume that the number of thermally accessible single-particle microstates is sufficiently large that the probability of multiple occupancy is negligible, so that $M(\mathbf{n})$ can be replaced by $N!$ in Eq. (8.13). Equations (8.13) and (8.14) then combine to imply

$$Z_N(\beta) = \frac{Z_1^N}{N!} \tag{8.15}$$

where

$$Z_1(\beta) = \sum_k \exp(-\beta\varepsilon_k) \tag{8.16}$$

is the single-particle partition function. The factor $1/N!$ in Eq. (8.15) is often regarded as a correction factor introduced by "corrected Boltzmann counting," but it appears automatically if the generic states are taken seriously from the outset as the true microstates of the system.

It is straightforward to evaluate $Z_1(\beta)$ in terms of the energies ε_{ijk} of the quantum states (i, j, k) of a single point particle of mass m confined to a rectangular box of dimensions (L_x, L_y, L_z). Those energies are derived in any quantum mechanics textbook, and are given by

$$\varepsilon_{ijk} = \frac{\pi^2\hbar^2}{2m}\left(\frac{i^2}{L_x^2} + \frac{j^2}{L_y^2} + \frac{k^2}{L_z^2}\right) \quad (i, j, k = 1, 2, \cdots) \tag{8.17}$$

where $\hbar$ is Planck's constant divided by 2π as usual, and each of the integers (i, j, k) varies from unity to infinity. Physical particles of interest (e.g., atoms and molecules) generally have internal states and energies as well (e.g., rotational, vibrational, and electronic). Such internal states are neglected here, but they can easily be included as described in standard textbooks. Combining Eqs. (8.16) and (8.17), we obtain

$$Z_1 = Z_x Z_y Z_z \tag{8.18}$$

where

$$Z_r \equiv \sum_{k=1}^{\infty} \exp(-\alpha_r k^2) \qquad (r = x, y, z) \tag{8.19}$$

and $\alpha_r \equiv \pi^2\hbar^2\beta/(2mL_r^2)$. It is easy to verify that $\alpha_r \ll 1$ for typical values of the parameters therein, except at extremely low temperatures. The number of terms of order unity in the summation in Eq. (8.19) is therefore of order $1/\sqrt{\alpha_r} \gg 1$, so $Z_r \gg 1$. Thus $Z_r - 1 \cong Z_r$, so the summation can be extended from $k = 0$ to ∞ with negligible error. Moreover, the summand $\exp(-\alpha_r k^2)$ is a very slowly decaying function of k, so the summation can be accurately approximated by an integral. Thus

$$Z_r \cong \int_0^{\infty} dk \, \exp(-\alpha_r k^2) = \frac{1}{2}\left(\frac{\pi}{\alpha_r}\right)^{1/2} = L_r \left(\frac{m}{2\pi\hbar^2\beta}\right)^{1/2} \tag{8.20}$$

Combining Eqs. (8.18) and (8.20), we obtain

$$Z_1(\beta) = V\left(\frac{m}{2\pi\hbar^2\beta}\right)^{3/2} \tag{8.21}$$

where $V = L_x L_y L_z$ is the volume of the box.

Equations (8.15) and (8.21) combine to determine the canonical partition function $Z_N(\beta)$ for the entire gas, from which its energy and entropy are easily obtained in just a few lines of algebra via Eqs. (5.37) and (5.38), or (7.6) and (7.13). All of its other thermodynamic functions and properties can then be derived from Eqs. (6.29), (6.41), (6.51), (6.54), and various thermodynamic identities. In particular, we leave it as an exercise for the reader to verify that the resulting pressure and energy are given by the familiar ideal-gas expressions $P = (N/V)\kappa T$ and $E = (3/2)N\kappa T$, and that the entropy is given by the famous Sackur-Tetrode equation [10].

8.6 Continuous Microstates

We now consider systems of N indistinguishable particles whose single-particle microstates are continuous rather than discrete, so that their multi-particle microstates are likewise continuous. The continuous single-particle microstates will be denoted by $\mathbf{x}$, so the state of particle α is denoted by $\mathbf{x}_\alpha$. The specific microstates of the N-particle system are then represented by the ordered N-tuples $\mathbf{x}^N \equiv (\mathbf{x}_1, \mathbf{x}_2, \cdots, \mathbf{x}_N)$. The configuration of N-particle systems is almost invariably described in terms of these specific states. This description is so familiar, natural, and convenient that one may tend to lose sight of the fact that the particle labels α and the specific states $\mathbf{x}^N$ are fictitious in the present context of indistinguishable particles. By definition, the real indistinguishable particles carry no labels whereby they might be distinguished.

The generic microstates of the N-particle system are defined by the unordered multisets $\{\mathbf{x}^N\} \equiv \{\mathbf{x}_1, \mathbf{x}_2, \cdots, \mathbf{x}_N\}$, which in the present context are more convenient to work with than occupation numbers. In contrast to its specific state, the generic state of the system is an objectively real feature which can be determined, at least in principle, simply by observing the system and making an unordered list of the N single-particle states $\mathbf{x}_\alpha$ occupied by its N particles, for which purpose the particles obviously need not be labeled. The number of specific states corresponding to each generic state is again simply a multinomial coefficient, but since the variables $\mathbf{x}_\alpha$ are continuous, specific states $(\mathbf{x}_1, \mathbf{x}_2, \cdots, \mathbf{x}_N)$ in which two or more of the values $\mathbf{x}_\alpha$ coincide constitute a set of measure zero in $\mathbf{x}^N$-space and can therefore be presumed to be infinitely improbable. The aforementioned multinomial coefficient then reduces to $N!$, which is therefore the number of specific states $\mathbf{x}^N$ corresponding to each generic state $\{\mathbf{x}^N\}$. Conversely, the fraction of the generic state $\{\mathbf{x}^N\}$ which is contributed by or can be attributed to each of its corresponding specific states $\mathbf{x}^N$ is $1/N!$.

As discussed in Sect. 2.6, however, for statistical purposes each single-particle state $\mathbf{x}$ cannot actually be regarded as a distinct microstate, because the number of such states in any finite region of $\mathbf{x}$-space, however small, is uncountably infinite. Thus it was necessary to introduce the density of states $\rho(\mathbf{x})$ in order to properly generalize the BGS entropy to systems with continuous states. The number of distinct single-particle states $\mathbf{x}$ within the differential element $d\mathbf{x}$ is then given by $\rho(\mathbf{x})\,d\mathbf{x}$. The number of specific states $(\mathbf{x}_1, \mathbf{x}_2, \cdots, \mathbf{x}_N)$ within each differential volume element $d\mathbf{x}^N \equiv d\mathbf{x}_1\,d\mathbf{x}_2 \cdots d\mathbf{x}_N$ in $\mathbf{x}^N$-space is therefore simply the product $\prod_\alpha \rho(\mathbf{x}_\alpha)\,d\mathbf{x}_\alpha = \rho_s(\mathbf{x}^N)\,d\mathbf{x}^N$, where

$$\rho_s(\mathbf{x}^N) \equiv \prod_{\alpha=1}^{N} \rho(\mathbf{x}_\alpha) \tag{8.22}$$

is the density of distinct specific states per unit volume in $\mathbf{x}^N$-space. Note that $\rho_s(\mathbf{x}^N)$ is invariant to permutations of its arguments $\mathbf{x}_\alpha$, and hence actually depends only on the generic state $\{\mathbf{x}^N\}$.

The integral $\int d\mathbf{x}^N \rho_s(\mathbf{x}^N)\,F(\mathbf{x}^N)$ represents the sum of the quantity $F(\mathbf{x}^N)$ over all distinct specific states. If $F(\mathbf{x}^N)$ depends only on the generic state $\{\mathbf{x}^N\}$ then its value for each such state appears $N!$ times in the sum. This implies that the sum of $F\{\mathbf{x}^N\}$ over all distinct *generic* states can be evaluated as the integral $\int d\mathbf{x}^N \rho_g(\mathbf{x}^N)\,F\{\mathbf{x}^N\}$, where

$$\rho_g(\mathbf{x}^N) \equiv \frac{1}{N!}\,\rho_s(\mathbf{x}^N) = \frac{1}{N!}\prod_{\alpha=1}^{N} \rho(\mathbf{x}_\alpha) \tag{8.23}$$

can be interpreted as the density of distinct generic states $\{\mathbf{x}^N\}$ per unit volume in the space of the specific states $\mathbf{x}^N$. It may at first seem incongruous to define such a hybrid object with one foot in each state space, so to speak, but this is precisely the tool required in the present context. The reason, as noted by Gibbs [17], is that integrating over generic states is difficult and cumbersome, whereas integrating over specific states is straightforward. Thus it is no longer convenient to sum directly over generic states, but since each such state corresponds to $N!$ specific states, sums over the former can easily be computed by summing over the latter and dividing the result by $N!$. This procedure is facilitated by defining $\rho_g(\mathbf{x}^N)$, in which the required division has already been performed in advance.

The joint probability density of specific states $\mathbf{x}^N$ per unit volume in $\mathbf{x}^N$-space is denoted by $p_s(\mathbf{x}^N)$, and is defined so that $p_s(\mathbf{x}^N)\, d\mathbf{x}^N$ is the probability that the specific state $\mathbf{x}^N$ of the system lies in the differential volume element $d\mathbf{x}^N$ at the point $\mathbf{x}^N$. The particles are identical except for their labels, so $p_s(\mathbf{x}^N)$ is invariant to permutations of its arguments $\mathbf{x}_\alpha$ and therefore actually depends only on the corresponding generic state $\{\mathbf{x}^N\}$. The number of distinct specific states in the volume element $d\mathbf{x}^N$ is $\rho_s(\mathbf{x}^N)\, d\mathbf{x}^N$, all of which have the same single-state probability $q_s(\mathbf{x}^N) = p_s(\mathbf{x}^N)/\rho_s(\mathbf{x}^N)$. Clearly $q_s(\mathbf{x}^N)$ also depends only on $\{\mathbf{x}^N\}$, which implies that all of the $N!$ distinct specific states $\mathbf{x}^N$ which correspond to a single distinct generic state $\{\mathbf{x}^N\}$ have that same probability $q_s(\mathbf{x}^N)$. The probability that the system occupies that single generic state is therefore

$$q_g(\mathbf{x}^N) = N!\, q_s(\mathbf{x}^N) = N!\, \frac{p_s(\mathbf{x}^N)}{\rho_s(\mathbf{x}^N)} = \frac{p_s(\mathbf{x}^N)}{\rho_g(\mathbf{x}^N)} \tag{8.24}$$

Since the generic states are the true microstates of the system, the BGS entropy must be evaluated in terms of them, and is therefore given by

$$S = -\int d\mathbf{x}^N\, \rho_g(\mathbf{x}^N)\, q_g(\mathbf{x}^N) \log q_g(\mathbf{x}^N) \tag{8.25}$$

Combining Eqs. (8.24) and (8.25), we obtain

$$S = -\int d\mathbf{x}^N\, p_s(\mathbf{x}^N) \log \frac{p_s(\mathbf{x}^N)}{\rho_g(\mathbf{x}^N)} = S_d - \log N! \tag{8.26}$$

where

$$S_d = -\int d\mathbf{x}^N\, p_s(\mathbf{x}^N) \log \frac{p_s(\mathbf{x}^N)}{\rho_s(\mathbf{x}^N)} \tag{8.27}$$

is again the entropy which the fictitious system of distinguishable particles would possess if it and they were real. The fact that $S \neq S_d$ is not in any sense paradoxical; it simply reflects the fact that the specific states are fictitious and are not the true microstates of the system. Equation (8.26) was previously obtained for particles with discrete microstates as Eq. (8.9). In contrast to the discrete case, however, Eqs. (8.26) and (8.27) now constitute the fundamental expression for the entropy of a system of indistinguishable particles with continuous microstates, since as discussed above it is inconvenient to directly integrate over the continuous generic microstates. The probability density $p_s(\mathbf{x}^N)$ itself is again to be determined by means of the relations derived in previous chapters, but for that purpose the multi-particle states of the entire system must of course be temporarily identified with its fictitious specific states $\mathbf{x}^N$ rather than its true generic states $\{\mathbf{x}^N\}$.

It should be noted, however, that the preceding development is not unconditionally self-consistent, because introducing the density of states $\rho(\mathbf{x})$ implies that multiply occupied single-particle states can no longer be considered infinitely improbable. The reason is that $\rho(\mathbf{x})$ is the number of distinct single-particle states per unit volume in $\mathbf{x}$-space, so $1/\rho(\mathbf{x})$ is conversely the volume in $\mathbf{x}$-space associated with each of those states. Thus two or more values of $\mathbf{x}$ which are so close together that their volumes intersect or overlap are effectively no longer distinct and should be regarded as a single multiply-occupied state. Of course, if $\rho(\mathbf{x})$ is sufficiently large then the volumes $1/\rho(\mathbf{x})$ will be so small that the probability of multiple occupancy remains negligible. However, this cannot be ascertained until $\rho(\mathbf{x})$ has been defined, so that the number $W_1(T)$ of thermally accessible single-particle microstates $\mathbf{x}$ with energies $\varepsilon(\mathbf{x}) \lesssim \kappa T$ can be evaluated in terms of $\rho(\mathbf{x})$. The condition for multiple occupancy to be negligible is then $W_1(T) \gg N$ as before, and the validity of Eqs. (8.24) and (8.26) is contingent on that condition.

The continuous multi-particle microstates of greatest interest in the present context are those which describe conservative Hamiltonian systems of classical particles. As discussed in Sect. 3.2.2, those microstates are normally defined in terms of the canonical coordinates and momenta of the particles. Thus the single-particle microstates $\mathbf{x}$ of such particles in three space dimensions are represented by the ordered sextuples $(\mathbf{q}, \mathbf{p})$, where $\mathbf{q}$ and $\mathbf{p}$ respectively represent ordered triplets of the canonical coordinates and momenta of the particle. The *specific* microstates $\mathbf{x}^N$ of a system of N identical three-dimensional classical particles will be represented by the ordered $6N$-tuples $(\mathbf{q}^N, \mathbf{p}^N) \equiv (\mathbf{q}_1, \cdots, \mathbf{q}_N, \mathbf{p}_1, \cdots, \mathbf{p}_N)$. It is conventional to refer to the state space of either single or multiple particles as a *phase space*. The canonical phase spaces $(\mathbf{q}, \mathbf{p})$ and $(\mathbf{q}^N, \mathbf{p}^N)$ are often referred to as μ-space and Γ-space, respectively. The volume elements in these spaces are denoted by $d\mathbf{x} = d\mathbf{q}\, d\mathbf{p}$ and $d\mathbf{x}^N = d\mathbf{q}^N d\mathbf{p}^N = \prod_\alpha d\mathbf{q}_\alpha d\mathbf{p}_\alpha$.

8.7 Classical Ideal Gases

It is of interest to compute the canonical partition function $Z_N(\beta)$ for a classical ideal gas and compare the result with Eqs. (8.15) and (8.21) for a quantum-mechanical ideal gas. The energy of a classical ideal gas is simply

$$E(\mathbf{q}^N, \mathbf{p}^N) = E(\mathbf{p}^N) = \sum_{\alpha=1}^{N} \varepsilon(\mathbf{p}_\alpha) \tag{8.28}$$

where $\varepsilon(\mathbf{p}) = |\mathbf{p}|^2/(2m)$ is the kinetic energy of a single particle with momentum $\mathbf{p}$. It is essential to remember that the states $(\mathbf{q}^N, \mathbf{p}^N)$ are fictitious specific microstates, whereas $Z_N(\beta)$ must be computed by summing the Boltzmann factor $\exp\{-\beta E(\mathbf{p}^N)\}$ over the true generic microstates. According to the discussion in the preceding section, this can be accomplished simply by multiplying $\exp\{-\beta E(\mathbf{p}^N)\}$ by $\rho_g(\mathbf{q}^N, \mathbf{p}^N)$ and integrating over the specific state space $(\mathbf{q}^N, \mathbf{p}^N)$.

Thus

$$Z_N(\beta) = \int d\mathbf{q}^N d\mathbf{p}^N \, \rho_g(\mathbf{q}^N, \mathbf{p}^N) \, \exp\{-\beta E(\mathbf{p}^N)\} \tag{8.29}$$

Combining Eqs. (8.23), (8.28), and (8.29), we obtain

$$Z_N(\beta) = \frac{1}{N!} \int d\mathbf{q}^N d\mathbf{p}^N \prod_{\alpha=1}^{N} \rho(\mathbf{q}_\alpha, \mathbf{p}_\alpha) \exp\{-\beta\varepsilon(\mathbf{p}_\alpha)\} \tag{8.30}$$

where $\rho(\mathbf{q}, \mathbf{p})$ is the density of states in the single-particle state space $(\mathbf{q}, \mathbf{p})$. The integral factors into a product of N identical single-particle integrals, with the result

$$Z_N(\beta) = \frac{Z_1^N}{N!} \tag{8.31}$$

where

$$Z_1(\beta) = \int d\mathbf{q} \, d\mathbf{p} \, \rho(\mathbf{q}, \mathbf{p}) \, \exp\{-\beta|\mathbf{p}|^2/(2m)\} \tag{8.32}$$

As discussed in Sect. 3.2.2, the natural definition of the single-particle density of states in the canonical variables $(\mathbf{q}, \mathbf{p})$ is $\rho(\mathbf{q}, \mathbf{p}) = \rho_0$ where ρ_0 is a constant. Equation (8.32) then reduces to

$$Z_1(\beta) = \rho_0 V \int d\mathbf{p} \, \exp\{-\beta|\mathbf{p}|^2/(2m)\} = \rho_0 V \left(\frac{2\pi m}{\beta}\right)^{3/2} \tag{8.33}$$

The value of ρ_0 is indeterminate classically, so it is natural to choose it in such a way that the classical and quantum-mechanical expressions for $Z_N(\beta)$ agree. Since Eqs. (8.15) and (8.31) are already identical, this can be accomplished by choosing ρ_0 so that Eqs. (8.21) and (8.33) for the single-particle partition function $Z_1(\beta)$ agree. We thereby obtain $\rho_0 = 1/h^3$, where $h = 2\pi\hbar$ is Planck's constant. Equation (8.23) for the density of generic states in specific state space then reduces to

$$\rho_g(\mathbf{q}^N, \mathbf{p}^N) = \frac{1}{N! \, h^{3N}} \tag{8.34}$$

Since $\rho(\mathbf{q}, \mathbf{p}) = \rho_0$ has now been determined, we can now confirm that multiple occupancy of single-particle states is indeed negligible as has been presumed. The number of thermally accessible single-particle states with energies $\lesssim \kappa T$ is simply $W_1(T) \sim \rho_0 \int d\mathbf{q}\, d\mathbf{p} = (1/h^3) \int d\mathbf{q}\, d\mathbf{p}$, where $\mathbf{q}$ is restricted to the volume V in position space and $\mathbf{p}$ is restricted to the spherical region $|\mathbf{p}|^2 \lesssim 2m\kappa T$ in momentum space. Thus $\int d\mathbf{q} = V$ and $\int d\mathbf{p} \sim (4\pi/3)(2m\kappa T)^{3/2}$, and we thereby obtain

$$W_1(T) \sim \frac{4\pi V}{3} \left(\frac{2m\kappa T}{h^2} \right)^{3/2} = \frac{4}{3\sqrt{\pi}} Z_1(\beta) \tag{8.35}$$

Thus $W_1(T) \sim Z_1(\beta)$, which in retrospect is obvious. For a hydrogen atom in a volume $V = 10^3$ cm^3 at $T = 300$ K, we find $Z_1 \cong 10^{27}$. This exceeds the number of atoms in that volume at atmospheric pressure by roughly a factor of 4×10^4, so $W_1 \gg N$ and multiple occupancy can be presumed negligible. In this context the condition $W_1 \gg N$ can be restated in the form $\Lambda^3 \ll V/N$, where $\Lambda \equiv h/(2\pi m\kappa T)^{1/2}$ is called the thermal de Broglie wavelength. The mean distance between particles is just $(V/N)^{1/3}$, so the condition $\Lambda^3 \ll V/N$ can be verbally restated as requiring the thermal de Broglie wavelength to be much smaller than the mean interparticle separation, which is the usual condition for quantum effects to be negligible.

We are now also at long last equipped to confirm the assertion made in Sect. 6.5 that the number of microstates $W(N, V, E)$ in an ideal gas is of the form $V^N C(N, E)$, because $W(N, V, E)$ can now be directly evaluated from Eq. (8.34):

$$W(N, V, E) = \int_{\mathbb{E}} d\mathbf{q}^N d\mathbf{p}^N \rho_g(\mathbf{q}^N, \mathbf{p}^N) = \frac{V^N}{N!\, h^{3N}} \int_{\mathbb{E}} d\mathbf{p}^N \tag{8.36}$$

where the subscript $\mathbb{E}$ indicates that the integrals are restricted to values of $(\mathbf{q}^N, \mathbf{p}^N)$ for which $|E(\mathbf{p}^N) - E| < (1/2)\Delta E$. Equation

(8.36) confirms the aforementioned assertion. The remaining restricted integral over momentum space can actually be evaluated analytically to obtain an explicit expression for $W(N, V, E)$ and thereby the entropy $S = \log W$ [10]. This is an interesting mathematical exercise, but it is much easier to evaluate the entropy and thermodynamic functions from the canonical or grand canonical partition function. Of course, when this is done the sharp constraint on the energy is effectively relaxed to a constraint on its mean value. As discussed in Chapter 6, however, the resulting energy fluctuations are negligible for thermodynamic systems, so the two procedures are equivalent. In ensemble language, this is referred to as the "equivalence of ensembles," but it must be remembered that it no longer obtains in small systems.

8.8 Bose-Einstein and Fermi-Dirac Statistics

Bose-Einstein (BE) and Fermi-Dirac (FD) statistics describe systems of indistinguishable particles in which multiple occupancy of single-particle states is either allowed (BE) or prohibited (FD). Thus the Pauli exclusion principle implies that fermions in quantum mechanics are described by FD statistics, whereas bosons are described by BE statistics. Contrary to the prevalent view, however, neither BE nor FD statistics is intrinsically quantum-mechanical in nature; they simply describe systems of indistinguishable particles with discrete states. This does not necessarily exclude classical systems, because the continuous classical phase space $\mathbf{x}^N$ can be discretized if desired, and indeed is already implicitly and effectively discretized by the density of states $\rho_g(\mathbf{x}^N)$, as discussed in Sect. 8.6. Moreover, the Pauli principle is not the only possible basis for prohibiting multiple occupancy of states. Some types of states are intrinsically restricted to single occupancy by their very nature; e.g., Chinese Checkers (which oddly originated in Germany). In thermodynamic systems, however, the distinctive features of BE and FD statistics are indeed associated with quantum mechanics, because they manifest themselves in situations where the number of thermally accessible single-particle states does not greatly exceed the number of particles; i.e., the thermal de Broglie wavelength Λ is comparable to or even larger than the mean interparticle separation $(V/N)^{1/3}$. Under these conditions multiple occupancy of single-particle states is expected to be significant unless it is prohibited; i.e., significant in the BE case but absent by executive fiat in the FD case.

According to Eqs. (5.36) and (7.5), the canonical partition function is given by $Z = \sum_k \exp(-\beta E_k)$, in which the summation extends over the accessible microstates k of the entire N-particle system. Those microstates are presumed to have definite energies E_k, which implies that in quantum mechanics they must be identified with the energy eigenstates $|k\rangle$ of the Hamiltonian operator $\mathcal{H}$; i.e., $\mathcal{H}|k\rangle = E_k|k\rangle$. The eigenstates $|k\rangle$ can always be orthonormalized, from which it follows that $\exp(-\beta E_k) = \langle k|\exp(-\beta\mathcal{H})|k\rangle$ and $Z = \text{Tr}\exp(-\beta\mathcal{H})$, where $\text{Tr}\,\mathcal{O} \equiv \sum_k \langle k|\mathcal{O}|k\rangle$ denotes the trace of the operator $\mathcal{O}$ in state space. In the preceding development, however, the true microstates of the system have been identified with the generic states $\mathbf{n}$, which correspond to the Fock states $|\mathbf{n}\rangle$ in the second quantization or occupation number representation [44, 45]. The latter identification is consistent only if the Fock states $|\mathbf{n}\rangle$ coincide with the energy eigenstates $|k\rangle$, so that $\mathcal{H}|\mathbf{n}\rangle = E(\mathbf{n})|\mathbf{n}\rangle$. It is straightforward to verify that such is indeed the case for systems of noninteracting particles, which confirms that Z can be written in the form of Eq. (8.13) for that case. However, if the N particles interact with one another then the Fock states are no longer eigenstates of $\mathcal{H}$ and no longer have definite energies $E(\mathbf{n})$. This complication can be formally circumvented by exploiting the fact that the trace has the same value for any orthonormal basis in state space. The Fock states constitute such a basis, so $Z = \text{Tr}\exp(-\beta\mathcal{H})$ can be evaluated in the basis $|\mathbf{n}\rangle$. Thus $Z = \sum_{\mathbf{n}} \langle\mathbf{n}|\exp(-\beta\mathcal{H})|\mathbf{n}\rangle$, in which the summation extends over the accessible Fock states of the N-particle system. The form of Eq. (8.13) can thereby be preserved even in quantum-mechanical systems of interacting particles by the mathematical expedient of defining $E(\mathbf{n}) \equiv -\kappa T \log\langle \mathbf{n}| \exp(-\beta \mathcal{H})|\mathbf{n}\rangle$, so that $\langle \mathbf{n}| \exp(-\beta \mathcal{H})|\mathbf{n}\rangle = \exp\{-\beta E(\mathbf{n})\}$. This expression for $E(\mathbf{n})$ simply reduces to the eigenvalues of $\mathcal{H}$ when the particles do not interact, so it remains applicable in that special case as well.

The net result of the preceding deliberations is that when $E(\mathbf{n})$ is interpreted in the generalized sense defined above, Eq. (8.13) for Z is generally applicable even for systems of interacting particles in quantum mechanics. Of course, the summations over $\mathbf{n}$ and $\mathbf{k}$ in Eq. (8.13) remain subject to the restriction that they extend only over the microstates accessible to the system. The only such restriction for BE statistics is that $|\mathbf{n}| = N$, which implies that $0 \leq n_k \leq N$. The canonical partition function for bosons is therefore given by

$$Z_{\mathrm{BE}} = \sum_{|\mathbf{n}|=N} \exp\{-\beta E(\mathbf{n})\} = \sum_{\mathbf{k}} \frac{\exp\{-\beta E(\mathbf{k})\}}{M(\mathbf{n})} \tag{8.37}$$

In the FD case, the occupation numbers n_k are subject to the further restriction $0 \leq n_k \leq 1$, which corresponds to the restriction $k_1 \neq k_2 \neq \cdots \neq k_N$ on the single-particle states k_α occupied by the N particles. A summation subject to these further FD restrictions will be indicated by a superscript $\oslash$. Thus the canonical partition function for fermions is given by

$$Z_{\mathrm{FD}} = \sum_{|\mathbf{n}|=N}^{\oslash} \exp\{-\beta E(\mathbf{n})\} \tag{8.38}$$

Since $M(\mathbf{n}) \leq N!$, Eq. (8.37) implies that

$$Z_{\mathrm{BE}} > \frac{1}{N!} Z_{\mathrm{DP}} \tag{8.39}$$

where

$$Z_{\mathrm{DP}} \equiv \sum_{\mathbf{k}} \exp\{-\beta E(\mathbf{k})\} \tag{8.40}$$

is the canonical partition function for the corresponding fictitious system of distinguishable particles. Equation (8.40) can be rewritten in the equivalent form

$$Z_{\mathrm{DP}} = \sum_{|\mathbf{n}|=N} M(\mathbf{n}) \exp\{-\beta E(\mathbf{n})\} > N!\, Z_{\mathrm{FD}} \tag{8.41}$$

in which the inequality is an immediate consequence of the fact that the summation extends over multiply occupied states in addition to the singly occupied states (for which $M(\mathbf{n}) = N!$) which contribute to Z_{FD}. Equations (8.39) and (8.41) combine to imply

$$Z_{\mathrm{FD}} < \frac{1}{N!} Z_{\mathrm{DP}} < Z_{\mathrm{BE}} \tag{8.42}$$

The inequalities in Eq. (8.42) show that if the difference between Z_{FD} and Z_{BE} is negligible then $Z_{\mathrm{FD}} \cong Z_{\mathrm{BE}} \cong (1/N!)\, Z_{\mathrm{DP}}$. As shown by Eqs. (8.37) and (8.38), the difference between Z_{FD} and Z_{BE} is simply that multiply occupied states are excluded from Z_{FD}. In situations where such states are already extremely improbable excluding them has no significant effect, in which case $Z_{\mathrm{FD}} \cong Z_{\mathrm{BE}} \cong (1/N!)\, Z_{\mathrm{DP}}$. As previously discussed, such is the case when the number of thermally accessible single-particle states greatly exceeds the number of particles, or equivalently when $\Lambda \ll (V/N)^{1/3}$. When that condition is satisfied there is no significant distinction between BE and FD statistics, and the canonical partition function for either can be conveniently computed simply by evaluating Z_{DP} and dividing by $N!$. Note that (a) the factor of $1/N!$ again emerges automatically from the formalism, and is not *ad hoc* in any sense; and (b) this result is not restricted to ideal gases but is equally valid for interacting particles, since the form of $E(\mathbf{n}) = E(\mathbf{k})$ has not been specified.

Particles which interact with and influence one another directly via interparticle forces and interaction energies are obviously not either dynamically or statistically independent. A noteworthy and counterintuitive feature of BE and FD statistics is that even ideal particles which do not directly interact with each other are nevertheless not in general statistically independent. This statistical dependence was illustrated in its simplest form in Sect. 8.1. In more general terms, it results from an interplay between indistinguishability and the EAPP hypothesis, which of course implicitly underlies the canonical and grand canonical probability distributions. It occurs in both FD and BE statistics, and is often interpreted in terms of an effective interparticle attraction (BE) or repulsion (FD). However, this language is metaphorical and should not be taken too seriously; the statistical dependence is neither the cause nor the consequence of any physical force. In the BE case, it is analogous to the fact that if twenty antisocial people check into a hotel with only ten rooms, some of them must perforce share a room, thereby creating the illusion that those who do are gregarious. In the FD case, occupied states cannot accept any further particles, thereby creating the illusion that the particles which occupy them are repelling the other particles.

In the remainder of this section we specialize to systems in which direct interparticle interactions are negligible, so that the energy of the system is again given by Eq. (8.14), which we rewrite here for convenience:

$$E(\mathbf{n}) = \sum_k n_k \, \varepsilon_k = E(\mathbf{k}) = \sum_{\alpha=1}^{N} \varepsilon(k_\alpha) \tag{8.43}$$

The presence or absence of statistical dependence can be ascertained by means of the relations derived in Sect. 8.4. For this purpose we must examine the canonical probability distribution of Eqs. (5.35) and (5.36), or (7.4) and (7.5), which in the present context takes the form

$$p_g(\mathbf{n}) = (1/Z_N) \exp\{-\beta E(\mathbf{n})\} \tag{8.44}$$

Equation (8.44) applies to both BE and FD statistics, with the understanding that the latter case implies the constraints $0 \leq n_k \leq 1$, so values of $\mathbf{n}$ inconsistent with those constraints do not represent accessible states of the system. We also require the probability p_k that an arbitrary particle in the system occupies state k, which as discussed in Sect. 7.1 given by

$$p_k = (1/Z_1) \exp(-\beta \varepsilon_k) \tag{8.45}$$

where $Z_1(\beta) = \sum_k \exp(-\beta\varepsilon_k)$. This is a general result for a single particle interacting weakly with a heat bath, which is therefore valid for both Bose-Einstein and Fermi-Dirac statistics and is not restricted to statistically independent particles. We can now subject Eq. (8.44) to the test for statistical independence provided by Eq. (8.12). Combining Eqs. (8.12), (8.43), and (8.45), we obtain, after a little algebra,

$$p_g(\mathbf{n}) = \frac{M(\mathbf{n})}{Z_{\mathrm{DP}}} \exp\{-\beta E(\mathbf{n})\} \tag{8.46}$$

This agrees with the true probability distribution of Eq. (8.44) for either BE or FD statistics only in the classical limit when $\Lambda \ll (V/N)^{1/3}$, in which $M(\mathbf{n})$ can be replaced by $N!$ and $Z_{\mathrm{BE}} \cong Z_{\mathrm{FD}} \cong Z_{\mathrm{DP}}/N!$. The particles are therefore statistically dependent (i.e., correlated) in both BE and FD statistics except in the classical limit.

Even for systems of particles which do not interact directly, the actual evaluation of Z_{BE} and Z_{FD} is difficult because of the constraint $|\mathbf{n}| = N$. In contrast, the grand canonical partition functions are easy to evaluate, so the remainer of the discussion is focused on the grand canonical probability distributions, partition functions, and mean occupation numbers for BE and FD statistics. The grand canonical probability distribution is given by Eqs. (5.42) and (5.43), or Eqs. (7.17) and (7.18), in which the microstates in the general formalism must again be identified with the generic microstates $\mathbf{n}$. Thus

$$p(\mathbf{n}) = (1/\Xi)\, \lambda^N \exp\{-\beta E(\mathbf{n})\} \tag{8.47}$$

where $\lambda = \exp(\beta\mu)$ as usual, and

$$\Xi(\beta, \lambda) = \sum_{\mathbf{n}} \lambda^N \exp\{-\beta E(\mathbf{n})\} \tag{8.48}$$

is the grand canonical partition function, in which the summation over $\mathbf{n}$ is now no longer subject to the restriction $|\mathbf{n}| = N$ and thereby automatically includes a summation over all values of $N = \sum_k n_k$.

Combining Eqs. (8.43), (8.47), and (8.48), we obtain

$$\Xi = \prod_k \Xi_k \tag{8.49}$$

where

$$\Xi_k \equiv \sum_n \lambda^n \exp(-\beta n \varepsilon_k) \tag{8.50}$$

and

$$p(\mathbf{n}) = \prod_k (1/\Xi_k)\, \lambda^{n_k} \exp(-\beta n_k \varepsilon_k) = \prod_k p_k(n_k) \tag{8.51}$$

where

$$p_k(n) \equiv (1/\Xi_k)\, \lambda^n \exp(-\beta n \varepsilon_k) \tag{8.52}$$

is the probability that precisely n particles occupy state k. Clearly $\sum_n p_k(n) = 1$, so $p_k(n)$ is properly normalized as it should be. The factorization of $p(\mathbf{n})$ in Eq. (8.51) implies that the occupation numbers n_k are statistically independent random variables, whereas in contrast the particles themselves are not, as we have seen. The mean or average occupation number of state k is given by

$$\bar{n}_k = \sum_n p_k(n)\, n = (1/\Xi_k) \sum_n n\, \lambda^n \exp(-\beta n \varepsilon_k) \tag{8.53}$$

Since $n\lambda^n = \lambda(d\lambda^n/d\lambda)$, $\bar{n}_k$ can be reexpressed in the more convenient form

$$\bar{n}_k = \frac{\lambda}{\Xi} \left(\frac{\partial\, \Xi_k}{\partial \lambda} \right) \tag{8.54}$$

The preceding formulae apply to both BE and FD statistics. The distinction between the two is that multiple occupancy of the states k (i.e., values of $n_k > 1$) are allowed in the former but not in the latter. The summation in Eq. (8.50) therefore extends over the range $0 \leq n < \infty$ in BE statistics, but only over the values $n = 0{,}1$ in FD statistics. In both cases the series can easily be summed analytically. In the BE case it is merely the simple geometric series $1/(1-x) = 1 + x + x^2 + \cdots$, and we thereby obtain

$$\Xi_k = \frac{1}{1 - \lambda \exp(-\beta\varepsilon_k)} \tag{8.55}$$

The FD case is even easier, since the series contains only two terms:

$$\Xi_k = 1 + \lambda \exp(-\beta\varepsilon_k) \tag{8.56}$$

The corresponding results for the mean occupation numbers are easily obtained from Eq. (8.54), and are given by

$$\bar{n}_k = \frac{1}{\lambda^{-1} \exp(\beta\varepsilon_k) \mp 1} \tag{8.57}$$

for the Bose-Einstein and Fermi-Dirac cases, respectively. In the latter case Eq. (8.53) reduces to $\bar{n}_k = p_k(1)$, so the probability that state k is occupied is simply the mean value of its occupation number.

It is instructive to examine the asymptotic behavior of the above relations in the classical limit $\Lambda \ll (V/N)^{1/3}$. It is clear from the preceding discussion and development that in the classical limit $\bar{n}_k \ll 1$ in both BE and FD statistics. It then follows from Eq. (8.57) that

$$\lambda \exp(-\beta\varepsilon_k) \ll 1 \tag{8.58}$$

in the classical limit, in which Eq. (8.57) therefore reduces to

$$\bar{n}_k = \lambda \exp(-\beta\varepsilon_k) \tag{8.59}$$

in both BE and FD statistics. Summing Eq. (8.59) over k, we obtain

$$\bar{N} \equiv \sum_k \bar{n}_k = \lambda Z_1(\beta) \tag{8.60}$$

where $Z_1(\beta) = \sum_k \exp(-\beta\varepsilon_k)$ is the single-particle partition function. Combining Eqs. (8.45), (8.59), and (8.60), we obtain the familiar single-particle Boltzmann distribution

$$\bar{n}_k/\bar{N} = (1/Z_1) \exp(-\beta\varepsilon_k) = p_k \tag{8.61}$$

Equation (8.58) further implies that Eqs. (8.55) and (8.56) differ negligibly in the classical limit, so that $\log \Xi_k \cong \lambda \exp(-\beta\varepsilon_k)$ and $\log \Xi = \sum_k \log \Xi_k = \lambda Z_1$. We thereby obtain

$$\Xi = \exp(\lambda Z_1) = \sum_{N=0}^{\infty} \lambda^N \frac{Z_1^N}{N!} \tag{8.62}$$

Comparison with Eq. (5.50) then shows that $Z_N = Z_1^N/N!$, in agreement with Eq. (8.15) of Sect. 8.5, which was derived by assuming that multiple occupancy was permitted but negligible. Those are precisely the conditions under which the particles would be expected to be statistically independent, which was confirmed by the consistency of Eqs. (8.44) and (8.46) in the classical limit.

Chapter 9

IDENTICAL PARTICLES ARE INHERENTLY INDISTINGUISHABLE

The preceding development has been specifically restricted to systems of identical indistinguishable particles, but no further conditions were imposed on the particles or their microstates. The relations derived in Chapter 8 are therefore equally applicable to systems of identical particles in both classical and quantum mechanics, *provided the particles are regarded as indistinguishable*. As discussed in the Preface and Introduction, it is our firm conviction that identical particles, by their very nature, are indeed inherently and unconditionally indistinguishable. If that proposition is accepted, the relations of Chapter 8 combine with the results of previous chapters to provide a unified general statistical formalism which applies to both classical and quantum particles. However, the appealing simplicity of this picture is undermined by the widespread misconception that identical classical particles are distinguishable, at least in principle, and must be treated as such. In this section we discuss the apparent origins and rationale for this unfortunate misconception in an attempt to reconstruct and rectify the fallacious notions on which it is based.

The assertion that identical particles of any type, classical or otherwise, are somehow distinguishable might seem oxymoronic to those unfamiliar with the subject matter. The terms "identical" and "indistinguishable" are very nearly synonymous in dictionaries and common usage, and to the extent that a distinction can be drawn between them it would seem to operate in the opposite direction. One can readily envision particles which are indistinguishable for all practical purposes even though they are not absolutely identical on a microscopic or molecular level; e.g., the two "indistinguishable" coins in Sect. 8.1.2. However, it is more difficult to imagine how one might operationally distinguish between two particles which are literally and absolutely identical in all respects down to the smallest microscopic or molecular detail; e.g., two hydrogen atoms. In what follows we critically analyze some possible ways whereby such a distinction might be effected in classical mechanics, and identify the flaws which make them untenable. The issues involved seem more conceptual than linguistic, so it hardly seems profitable to dwell any further on the semantics of the situation, especially since all definitions inevitably possess some residual degree of ambiguity (cf. Swendsen [8]).

Since the point has sometimes occasioned confusion, we reemphasize that in contrast to the identical particles themselves, the states they occupy must always be labeled and are consequently always distinguishable. This viewpoint is fully consistent with quantum mechanics, where the states are labeled by their quantum numbers but the indistinguishable identical particles are not. What is less widely appreciated is that these concepts are more generally applicable and actually have nothing to do with quantum mechanics per se.

9.1 Historical Remarks

Prior to the advent of statistical mechanics and quantum mechanics the distinguishability of identical particles was not an issue, because the dynamical evolution of a system of N particles in classical mechanics is indifferent to whether or not they are labeled. The states occupied by the particles are denoted by the variables $\mathbf{x}_\alpha$ ($\alpha = 1, \cdots, N$) in either case, but as discussed in Sect. 8.3 the interpretation of those variables differs in the two cases. If the particles are labeled then $\mathbf{x}_\alpha$ denotes the state occupied by particle α, whereas if the particles are identical and unlabeled the variables $\mathbf{x}_\alpha$ merely denote the arbitrarily labeled states which are occupied by the particles, regardless of the particular particles which occupy them. In either case, the time evolution of the variables $\mathbf{x}_\alpha(t)$ is determined by the same classical equations of motion, so even if the particles are identical and indistinguishable their dynamical time evolution can be computed as though they were labeled and distinguishable. When this is done, it is mathematically convenient to describe the entire N-particle system in terms of its specific states $\mathbf{x}^N = (\mathbf{x}_1, \cdots, \mathbf{x}_N)$. The fact that those states become fictitious for identical particles is therefore immaterial for dynamical purposes alone, where it has no physical consequences.

The development of the atomic theory of matter in the nineteenth century led naturally to the application of classical mechanics to atoms and molecules. In contrast to macroscopic particles, atoms and molecules of the same type cannot be physically labeled and really are absolutely identical in all respects. The only obvious sense in which such particles can be regarded as labeled is in the fictitious *gedanken* sense discussed in Sect. 8.2. The specific states $\mathbf{x}^N$ defined in terms of those labels are then likewise fictitious, but this is easily disregarded or overlooked because it has no purely dynamical consequences. As Gibbs observed, however, it has profound *statistical* implications, but these were not recognized until they manifested themselves in the formulation of statistical mechanics. A careful reading of Gibbs [17] leaves little if any doubt that he clearly understood the essential points discussed in Sect. 8.6, namely that (a) identical particles are inherently indistinguishable; (b) the true microstates of a system of such particles are therefore its generic states $\{\mathbf{x}^N\}$ rather than its specific states $\mathbf{x}^N$; (c) it is difficult to directly sum over the generic states, but easy to sum over the fictitious specific states; and (d) since multiple occupancy of continuous states is extremely improbable, the generic states can be correctly enumerated simply by summing over the specific states and dividing by $N!$. Unfortunately, the understated and somewhat tentative manner in which Gibbs expressed these important insights was ultimately unsuccessful in cogently communicating them to a critical mass of his contemporaries and successors. These insights were consequently not fully appreciated at the time, and indeed their applicability to identical classical particles remains widely unappreciated over a century later. Thus it is still not uncommon to see the Gibbs correction factor of $1/N!$ inaccurately denigrated by such terms as "ad hoc" or even "mystical," whereas in reality it is meticulously well founded as discussed in Sect. 8.6.

The inherent indistinguishability of identical particles was belatedly forced into the public consciousness by quantum mechanics. However, it was still widely believed that classical particles are inherently distinguishable, so indistinguishability was not yet understood as a more general principle and was consequently misinterpreted as one of the many profound differences between classical and quantum mechanics. This non sequitur launched the widespread misconception that indistinguishability is an intrinsically quantum concept, which persists to the present day. This misconception may have been reinforced by the fact that the states of classical particles are defined by simultaneously specifying their positions and momenta, which is forbidden in quantum mechanics by the uncertainty principle. This is logically unrelated to indistinguishability; it merely reflects the fundamental differences between the classical and quantum-mechanical state descriptions, but it may predispose the mind to entertain the erroneous notion that those differences include indistinguishability as well.

9.2 Distinguishability by Particle Interchange

Some authors claim that identical classical particles are distinguishable because interchanging them would produce an observable change in the physical state of the system as represented by its phase point $\mathbf{x}^N = (\mathbf{x}_1, \cdots, \mathbf{x}^N)$, whereas if they were indistinguishable such an interchange should have no detectable effect. This argument is sometimes stated in terms of the observation that interchanging particles would produce an unphysical discontinuity in the time-dependent trajectory $\mathbf{x}^N(t)$ which the system traces out in $\mathbf{x}^N$-space as determined by the classical equations of motion. However, such arguments are fallacious because the state space to which they refer is the space of *specific* states $\mathbf{x}^N$, which like the particle labels α used to define it is fictitious unless the identical particles can somehow be labeled in a physically observable way. Fictitious labels introduced for mathematical convenience are not physically observable and therefore do not qualify as a distinguishing characteristic of identical particles. If the particles possess no such characteristics, the true microstates of the system are its generic states $\{\mathbf{x}^N\}$, which are unordered multisets and are clearly unaffected by interchanging particles.

Fallacious arguments of the present type are based upon a failure to recognize that the specific states $\mathbf{x}^N$ do not define or represent the true states of a system of identical particles. Misinterpreting the specific states $\mathbf{x}^N$ as real rather than fictitious is a very natural and highly insidious error, because the variables $\mathbf{x}^N$ are already familiar from classical mechanics, where they provide a natural, convenient, and economical description of the time evolution of particulate systems. Moreover, that description is entirely legitimate for dynamical purposes alone even when the particles are identical, as discussed in Sect. 9.1. In other words, the variables $\mathbf{x}^N$ provide a valid description of the system for dynamical purposes, but not for statistical purposes if the particles are identical.

The statistical aspects of particulate systems are much more clearly exhibited and easily understood when their states are visualized as a cloud of N points in the single-particle state space $\mathbf{x}$ rather than a single point in the space of the N-particle specific states $\mathbf{x}^N$ [13]. This eliminates not only the danger but even the possibility of misinterpreting the significance of the specific states $\mathbf{x}^N$. The single-particle states occupied by a system of N indistinguishable particles, and thereby the generic state of the system as a whole, are represented by a cloud of unlabeled identical points in $\mathbf{x}$-space. The appearance of this cloud is clearly unchanged by interchanging particles, which makes it clear at a glance that the true states of such a system are its generic states. It is equally clear that the specific state of such a system is fictitious, because it cannot be determined by observing or inspecting the cloud. Conversely, if the particles are distinguishable then each point in the cloud can be labeled by the label (or labels) α of the particle (or particles) occupying that state. Such a cloud simultaneously shows which state is occupied by each of the particles, and thereby represents the specific state of the entire system.

9.3 Distinguishability by Implicit Labeling

Even those who contend that identical atoms or molecules are distinguishable would presumably not go so far as to claim that they can be physically labeled. And even if they could be, they would thereby become distinguishable but would no longer be identical. How then might identical particles conceivably be distinguished without physically labeling them? It might at first appear that the fictitious labels $\alpha = 1, \cdots, N$ introduced in Sect. 8.2 would suffice for this purpose. However, this idea is untenable because those labels are affixed to fictitious particles in a fictitious system, not the actual physical system of interest. Moreover, even if those labels could somehow be regarded as "associated" with the real particles in some nebulous sense, they are obviously not intrinsic or objective features of the particles which could be observed by inspection. Those fictitious labels therefore cannot be used to physically distinguish between the particles, which of course is necessary to unambiguously define or determine which particular particles occupy which single-particle states.

The possibility might also be contemplated of physically distinguishing between the real particles by devising some way of labeling them implicitly or indirectly; i.e., by proxy, as it were. In particular, one might wonder if they could be indirectly labeled in terms of the states they occupy. This idea superficially seems consistent with the use of the variables $\mathbf{x}^N = (\mathbf{x}_1, \cdots, \mathbf{x}_N)$ to describe the dynamics of a system of identical particles in classical mechanics, as discussed in Sect. 9.1. In that context the integers $\alpha = 1, \cdots, N$ are not particle labels and simply serve to label the occupied states $\mathbf{x}_\alpha$, but it is customary to loosely refer to the particle occupying state $\mathbf{x}_\alpha$ as particle α, and thereby to $\mathbf{x}_\alpha$ itself as the state occupied by particle α and $\mathbf{x}_\alpha(t)$ as the trajectory of particle α. Indeed, whichever particular particle occupies the state $\mathbf{x}_\alpha(0)$ at time $t = 0$ remains on the trajectory $\mathbf{x}_\alpha(t)$ for all t, so that particle might appear to be uniquely identified and hence implicitly labeled by its trajectory $\mathbf{x}_\alpha(t)$, or equivalently by its initial state $\mathbf{x}_\alpha(0)$ or the index α. This appearance is entirely illusory, however, because the particle already occupies the initial state $\mathbf{x}_\alpha(0)$ before that state is arbitrarily labeled, and the indeterminacy as to which particle occupies it is not removed by labeling it. Once the occupied states have been arbitrarily labeled by the integers α, $\mathbf{x}_\alpha$ simply represents one of those states and does not in any sense specify or determine which particle occupies it. Indeed, $\mathbf{x}_\alpha$ could not possibly convey such information because the particles themselves bear no labels whereby they could be identified. Similarly, $\mathbf{x}_\alpha(t)$ simply represents the trajectory of whichever particle occupied the state $\mathbf{x}_\alpha(0)$ at the initial time $t = 0$, and nothing more.

In short, when the particles are identical the common custom or convention of referring to $\mathbf{x}_\alpha$ as the state occupied by particle α is merely a convenient but imprecise linguistic abbreviation which cannot be interpreted literally. This loose terminology is useful in classical mechanics but hazardous for statistical purposes, because it creates the danger of misinterpreting the labels α of the occupied states as particle labels and thereby erroneously inferring that identical particles are distinguishable. We therefore conclude that identical particles cannot in fact be implicitly or indirectly labeled by proxy in terms of the states they occupy.

9.4 Distinguishability by Surveillance

The states of classical particles are specified in terms of their positions and momenta, which in principle can be simultaneously measured or observed without perturbing the system, in contrast to quantum particles. On this basis, some authors maintain that identical classical particles could in principle be distinguished without physically labeling them simply by keeping them under close surveillance; i.e., by continuously observing each and every individual particle and the trajectory it follows as the N-particle system evolves in time. However, in a macroscopic system of $N \sim 10^{23}$ particles this would require retaining the services of $\sim 10^{23}$ infallable Maxwellian demons, each of which is assigned the task of keeping a particular particle under constant surveillance. The demons themselves are presumed to be distinguishable and labeled, so employing them to distinguish between the particles essentially constitutes another variant of implicit labeling, wherein each particle is indirectly labeled by the label of the demon assigned to observe it. Aside from its fanciful impracticality, the essential conceptual fallacy with this proposal is that the demons are fictitious, so their labels are likewise fictitious. As discussed in the previous section, fictitious labels cannot be used to physically distinguish between real particles, and are therefore useless for defining or determining which particles occupy which states.

9.5 Discussion

Several possible approaches and methods for distinguishing between identical classical particles have been discussed, none of which survives close scrutiny. Of course, the inherent indistinguishability of identical particles in quantum mechanics has long been accepted. The purpose of the preceding discussion was to elaborate in further detail the basis and rationale for our contention that the indistinguishability of identical particles is in reality a more general and fundamental principle which is equally applicable in classical mechanics. Once this principle is accepted, the conceptual distinction between classical and quantum statistical mechanics disappears, and both are unified within the same general framework.

BIBLIOGRAPHY

F. C. Andrews, *Equilibrium Statistical Mechanics* (Wiley, New York, 1963).

P. W. Atkins, *The 2nd Law: Energy, Chaos, and Form* (Scientific American/Freeman, New York, 1994).

P. Attard, *Thermodynamics and Statistical Mechanics: Equilibrium by Entropy Maximisation* (Academic/Elsevier, London/San Diego, 2002.

R. Baierlein, *Thermal Physics* (Cambridge U.P., Cambridge, 1999).

H. A. Bent, *The Second Law* (Oxford U.P., New York, 1965).

M. Born, *Natural Philosophy of Cause and Chance* (Dover, New York, 1964).

L. Brillouin, *Science and Information Theory,* 2nd ed. (Academic, New York, 1962).

L. Brillouin, *Scientific Uncertainty, and Information* (Academic, New York, 1964).

D. Chandler, *Introduction to Modern Statistical Mechanics* (Oxford U.P., Oxford, 1987).

R. T. Cox, *The Algebra of Probable Inference* (Johns Hopkins, Baltimore, 1961).

N. Davidson, *Statistical Mechanics* (McGraw-Hill, New York, 1962).

S. R. de Groot and P. Mazur, *Non-Equilibrium Thermodynamics* (North-Holland, Amsterdam, 1969; Dover, New York, 1984).

P. S. Epstein, *Textbook of Thermodynamics* (Wiley, New York, 1937).

J. D. Fast, *Entropy* (Philips, Eindhoven, 1962).

E. Fermi, *Thermodynamics* (Dover, New York, 1956).

D. D. Fitts, *Nonequilibrium Thermodynamics* (McGraw-Hill, New York, 1962).

R. H. Fowler, *Statistical Mechanics* (Cambridge U.P., Cambridge, 1966).

R. H. Fowler and E. A Guggenheim, *Statistical Thermodynamics: A Version of Statistical Mechanics for Students of Physics and Chemistry* (Cambridge U.P., New York, 1965).

H. L. Friedman, *A Course in Statistical Mechanics* (Prentice-Hall, Englewood Cliffs, NJ, 1985).

J. W. Gibbs, *The Scientific Papers of J. Willard Gibbs, Vol. I: Thermodynamics* (Dover, New York, 1961).

D. T. Gillespie, *Markov Processes: An Introduction for Physical Scientists* (Academic, San Diego, 1992).

B. V. Gnedenko, *The Theory of Probability,* 4th ed. (Chelsea, New York, 1968).

B. V. Gnedenko and A. Ya. Khinchin, *An Elementary Introduction to the Theory of Probability* (Dover, New York, 1962).

D. L. Goodstein, *States of Matter* (Dover, New York, 1985).

H. Gould and J. Tobochnik, *Statistical and Thermal Physics: With Computer Applications* (Princeton U.P., Princeton, 2010).

E. A. Guggenheim, *Thermodynamics: An Advanced Treatment for Chemists and Physicists* (North-Holland, Amsterdam, 1967).

G. N. Hatsopoulos and J. H. Keenan, *Principles of General Thermodynamics* (Wiley, New York, 1965).

T. L. Hill, *An Introduction to Statistical Thermodynamics* (Addison-Wesley, Reading, MA, 1960).

T. L. Hill, *Thermodynamics of Small Systems* (Dover, New York, 1994).

J. O. Hirschfelder, C. F. Curtiss, and R. B. Bird, *Molecular Theory of Gases and Liquids* (Wiley, New York, 1954).

K. Huang, *Statistical Mechanics,* 2nd ed. (Wiley, New York, 1987).

E. A. Jackson, *Equilibrium Statistical Mechanics* (Dover, New York, 2000).

H. Jeffreys, *Theory of Probability,* 3rd ed. (Oxford U.P., Oxford, 1961).

G. Jumarie, *Relative Information: Theories and Applications* (Springer-Verlag, Berlin, 1990).

J. Kestin, *A Course in Thermodynamics* (Ginn/Blaisdell, Waltham, MA, 1966).

J. M. Keynes, *A Treatise on Probability* (Harper & Row, New York, 1962).

A. I. Khinchin, *Mathematical Foundations of Information Theory* (Dover, New York, 1957).

A. I. Khinchin, *Mathematical Foundations of Statistical Mechanics* (Dover, New York, 1949).

J. G. Kirkwood and I. Oppenheim, *Chemical Thermodynamics* (McGraw-Hill, New York, 1961).

C. Kittel and H. Kroemer, *Thermal Physics,* 2nd ed. (Freeman, New York, 1980).

R. Kubo, *Thermodynamics: An Advanced Course with Problems and Solutions* (North-Holland, Amsterdam, 1968).

R. Kubo, *Statistical Mechanics: An Advanced Course with Problems and Solutions* (North-Holland, Amsterdam, 1971).

P. T. Landsberg, *Thermodynamics* (Interscience, New York, 1961).

P. T. Landsberg, *Thermodynamics and Statistical Mechanics* (Oxford U.P., Oxford, 1978).

P.-S. Laplace, *A Philosophical Essay on Probabilities* (Dover, New York, 1995).

G. N. Lewis and M. Randall, *Thermodynamics,* 2nd ed., rev. by K. S. Pitzer and L. Brewer (McGraw-Hill, New York, 1961).

E. M. Lifshitz and L. P. Pitaevskii, *Statistical Physics,* 3rd ed., Parts 1 and 2 (Pergamon, Oxford, 1980).

R. B. Lindsay, *Introduction to Physical Statistics* (Wiley, New York, 1941).

D. K. C. MacDonald, *Noise and Fluctuations: An Introduction* (Wiley, New York, 1962).

D. K. C. MacDonald, *Introductory Statistical Mechanics for Physicists* (Dover, New York, 2006).

J. E. Mayer and M. G. Mayer, *Statistical Mechanics,* 2nd ed. (Wiley, New York, 1977).

D. A. McQuarrie, *Statistical Mechanics* (Harper & Row, New York, 1976).

P. M. Morse, *Thermal Physics,* rev. ed. (Benjamin, New York, 1964).

J. R. Partington, *An Advanced Treatise on Physical Chemistry,* Vol. I (Wiley, New York, 1949).

R. K. Pathria and P. D. Beale, *Statistical Mechanics*, 3rd ed. (Elsevier/Butterworth-Heinemann/Academic, Amsterdam/Oxford/Cambridge, MA, 2011).

O. Penrose, *Foundations of Statistical Mechanics* (Pergamon, Oxford, 1970).

A. B. Pippard, *The Elements of Classical Thermodynamics* (Cambridge U.P., Cambridge, 1957).

M. Planck, *Treatise on Thermodynamics,* 3rd ed. (Dover, New York, undated).

L. E. Reichl, *A Modern Course in Statistical Mechanics,* 2nd ed. (Wiley, New York, 1998).

H. Reiss, *Methods of Thermodynamics* (Ginn/Blaisdell, New York, 1965).

A. Rényi, *Probability Theory* (Dover, New York, 2007).

P. C. Riedi, *Thermal Physics,* 2nd ed. (Oxford U.P.,Oxford, 1988).

A. Sommerfeld, *Thermodynamics and Statistical Mechanics* (Academic, New York, 1964).

R. E. Sonntag and G. J. Van Wylen, *Fundamentals of Statistical Thermodynamics* (Wiley, New York, 1966).

D. ter Haar, *Elements of Thermostatistics* Holt, Rinehart, and Winston, New York, 1966).

D. ter Haar, *Elements of Statistical Mechanics,* 3rd ed. (Butterworth-Heinemann, Oxford, 1995).

C. J. Thompson, *Mathematical Statistical Mechanics* (Macmillan, New York, 1972).

C. L. Tien and J. H. Lienhard, *Statistical Thermodynamics* (Holt, Rinehart, and Winston, New York, 1971).

L. Tisza, *Generalized Thermodynamics* (M.I.T. Press, Cambridge, MA, 1966).

M. Toda, R. Kubo, and N. Saitô, *Statistical Physics I: Equilibrium Statistical Mechanics* (Springer-Verlag, Berlin, 1983).

G. E. Uhlenbeck and G. W. Ford, *Lectures in Statistical Mechanics* (American Mathematical Society, Providence, 1963).

J. V. Uspensky, *Introduction to Mathematical Probability* (McGraw-Hill, New York, 1937).

N. G. van Kampen, *Stochastic Processes in Physics and Chemistry*, 2nd ed. (North-Holland, Amsterdam, 1992).

G. H. Wannier, *Statistical Physics* (Wiley, New York, 1966).

L. C. Woods, *Thermodynamics of Fluid Systems* (Oxford U.P., Oxford, 1975).

REFERENCES

[1] R. H. Swendsen, "Statistical mechanics of classical systems with distinguishable particles," *J. Stat. Phys.* **107**, 1143–1166 (2002).

[2] J. F. Nagle, "Regarding the entropy of distinguishable particles," *J. Stat. Phys.* **117**, 1047–1062 (2004).

[3] R. H. Swendson, "Gibbs' paradox and the definition of entropy," *Entropy* **10**, 15–18 (2008).

[4] J. F. Nagle, "In defense of Gibbs and the traditional definition of the entropy of distinguishable particles," *Entropy* **12**, 1936–1945 (2010).

[5] R. H. Swendsen, "How physicists disagree on the meaning of entropy," *Am. J. Phys.* **79**, 342–348 (2011).

[6] R. H. Swendsen, "Choosing a definition of entropy that works," *Founds. Phys.* **42**, 582–593 (2012).

[7] R. H. Swendsen, "Unnormalized probability: A different view of statistical mechanics," *Am. J. Phys.* **82**, 941–946 (2014).

[8] R. H. Swendsen, "The ambiguity of 'distinguishability' in statistical mechanics," *Am. J. Phys.* **83**, 545 (2015).

[9] G. S. Rushbrooke, *Introduction to Statistical Mechanics* (Oxford U.P., Oxford, 1949).

[10] R. Becker, *Theory of Heat,* 2nd ed. (Springer, New York, 1967).

[11] A. Katz, *Principles of Statistical Mechanics: The Information Theory Approach* (Freeman, San Francisco, 1967).

[12] D. Hestenes, "Entropy and indistinguishability," *Am. J. Phys.* **38**, 840–845 (1970).

[13] S. Fujita, *Statistical and Thermal Physics,* Part I (Krieger, Malabar, FL, 1986).

[14] S. Fujita, "On the indistinguishability of classical particles," *Founds. Phys.* **21**, 439–457 (1991).

[15] A. Bach, *Indistinguishable Classical Particles* (Springer, Berlin, 1997).

[16] S. Saunders, "On the explanation for quantum statistics," *Stud. Hist. Phil. Mod. Phys.* **37**, 192–211 (2006).

[17] J. W. Gibbs, *Elementary Principles in Statistical Mechanics* (Dover, New York, 1960).

[18] R. Balian, *From Microphysics to Macrophysics: Methods and Applications of Statistical Physics,* Vol. I (Springer-Verlag, Berlin, 1991).

[19] E. T. Jaynes, *Probability Theory* (Cambridge U.P., Cambridge, 2003).

[20] R. D. Levine and M. Tribus, eds., *The Maximum Entropy Formalism* (MIT Press, Cambridge, MA, 1979).

[21] E. T. Jaynes, *Papers on Probability, Statistics and Statistical Physics,* R. D. Rosenkrantz, ed. (Reidel, Dordrecht, 1983).

[22] J. D. Ramshaw, "Remarks on entropy and irreversibility in non-Hamiltonian systems," *Phys. Lett. A* **116**, 110–114 (1986).

[23] H. Grabert, R. Graham, and M. S. Green, "Fluctuations and nonlinear processes. II," *Phys. Rev. A* **21**, 2136–2146 (1980).

[24] E. Schrödinger, *Statistical Thermodynamics* (Cambridge U.P., Cambridge, 1964).

[25] J. D. Ramshaw, "Remarks on non-Hamiltonian statistical mechanics," *Europhys. Lett.* **59**, 319–323 (2002).

[26] R. C. Tolman, *The Principles of Statistical Mechanics* (Oxford U.P., Oxford, 1938).

[27] A. Ben-Naim, "Is mixing a thermodynamic process?" *Am. J. Phys.* **55**, 725–733 (1987).

[28] A. Ben-Naim, "On the So-Called Gibbs Paradox, and on the Real Paradox," *Entropy* **9**, 132–136 (2007).

[29] H. B. Callen, *Thermodynamics and an Introduction to Thermostatistics,* 2nd ed. (Wiley, New York, 1985).

[30] B. F. Schutz, *Geometrical Methods of Mathematical Physics* (Cambridge U.P., Cambridge, 1980).

[31] R. L. Bishop and S. I. Goldberg, *Tensor Analysis on Manifolds* (Dover, New York, 1980).

[32] M. Tribus, *Thermostatics and Thermodynamics* (Van Nostrand, Princeton, 1961).

[33] R. Baierlein, *Atoms and Information Theory* (Freeman, San Francisco, 1971).

[34] A. Hobson, *Concepts in Statistical Mechanics* (Gordon and Breach, New York, 1971).

[35] A. Ben-Naim, *A Farewell to Entropy: Statistical Thermodynamics Based on Information* (World Scientific, Singapore, 2008).

[36] E. T. Jaynes, "Gibbs vs Boltzmann Entropies," *Am. J. Phys.* **33**, 391–398 (1965).

[37] W. T. Grandy, Jr., *Entropy and the Time Evolution of Macroscopic Systems* (Oxford U.P., Oxford, 2008).

[38] R. P. Feynman, *Statistical Mechanics: A Set of Lectures* (Benjamin/Cummings, Reading, MA, 1972).

[39] T. L. Hill, *Statistical Mechanics* (McGraw-Hill, New York, 1956).

[40] E. L. Knuth, *Introduction to Statistical Thermodynamics* (McGraw-Hill, New York, 1966).

[41] M. P. Allen and D. J. Tildesley, *Computer Simulation of Liquids* (Oxford U.P., Oxford, 1987).

[42] C. Garrod, *Statistical Mechanics and Thermodynamics* Oxford U.P., New York, 1995).

[43] F. Reif, *Fundamentals of Statistical and Thermal Physics* (McGraw-Hill, New York, 1965).

[44] N. H. March, W. H. Young, and S. Sampanthar, *The Many-Body Problem in Quantum Mechanics* (Cambridge U.P., Cambridge, 1967).

[45] A. L. Fetter and J. D. Walecka, *Quantum Theory of Many-Particle Systems* (McGraw-Hill, New York, 1971).

INDEX

Printed in the United States
By Bookmasters

Printed in the United States
By Bookmasters